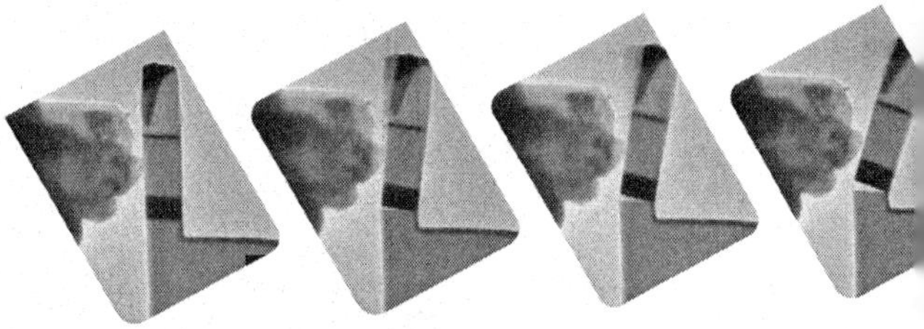

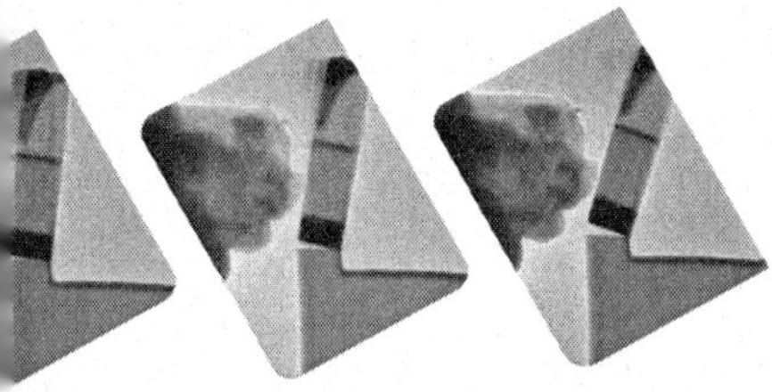

FRACTURE NANOMECHANICS

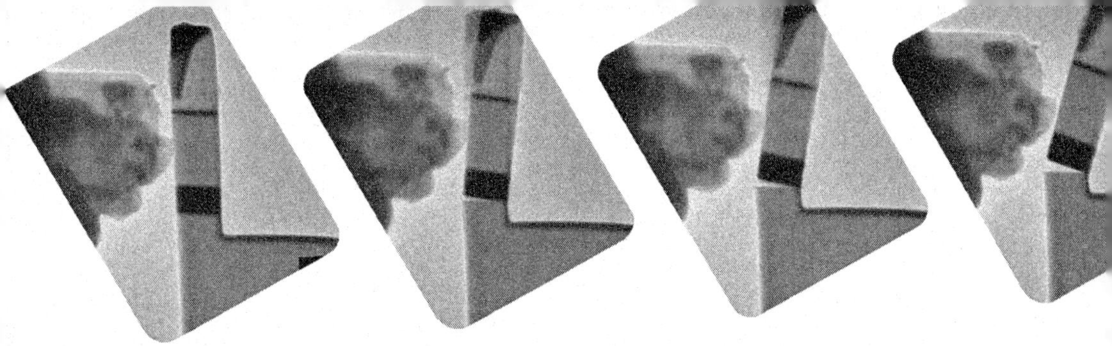

Takayuki Kitamura
Hiroyuki Hirakata
Takashi Sumigawa
Takahiro Shimada

FRACTURE NANOMECHANICS

PAN STANFORD PUBLISHING

Published by

Pan Stanford Publishing Pte. Ltd.
Penthouse Level, Suntec Tower 3
8 Temasek Boulevard
Singapore 038988

Email: editorial@panstanford.com
Web: www.panstanford.com

British Library Cataloguing-in-Publication Data
A catalogue record for this book is available from the British Library.

Fracture Nanomechanics

ISBN 978-981-4241-83-0 (Hardcover)
ISBN 978-981-4303-11-8 (eBook)

Printed in the USA

Contents

Chapter 1

INTRODUCTION

1.1 STRENGTH OF NANO-COMPONENTS

With recent developments in and miniaturization of electronic devices because of the increasing demand for high-density integration, the size of elements has been approaching a few nanometers. In an extreme, the atom-by-atom structure of atomic-scale elements, e.g., carbon nanotubes, comes in sight for future applications. Industry has intensely developed devices on the micrometer or nanometer scale that have multifarious functions—namely the micro-electro-mechanical system (MEMS) and nano-electro-mechanical system (NEMS), including small sensors and actuators. The stress applied to a nano-component stems from various sources, such as mechanical loading (e.g., polishing), residual stress, and thermal elongation mismatch in processing as well as service. Hence, to ensure reliability, nano-components should be carefully designed on the basis of mechanical conditions. Because they are composed of various materials with different nano-size geometries, the stress concentration originating from their shapes plays a critical role in the fracture mechanisms of nano-components. Moreover, to realize high-density integration of the components in devices, small materials with different mechanical properties (e.g., elastic modulus) are adhered each other without an interlayer or with an ultrathin one. Since the mismatch of deformation often

Fracture Nanomechanics
Edited by Takayuki Kitamura, Hiroyuki Hirakata, Takashi Sumigawa and Takahiro Shimada

www.panstanford.com

brings about inner stress, its interface is the most critical site in terms of a fracture.

In a complex process from a viewpoint of mechanics, such as the nano-imprint technique, where a thin film is removed along the interface (intentional delamination) after it is formed on a substrate, the interface strength is of central importance, and constitutive understanding of the fracture phenomenon is inevitable for its precise process management. The key issue in the technology is to spontaneously control the interface strength; adhesion and exfoliation. This clearly points out that we need fundamental knowledge of "fracture" not only for prevention of device malfunction but also for fabrication in the future technology of nano-components, namely from fracture prevention to fracture design. For this purpose, it is essential to understand both the mechanism and mechanics of material strength, such as deformation and failure characteristics of nanometer-scale materials.

1.2 FRACTURE NANO-MECHANICS IN STRUCTURE

We cannot understand the strength of nano-components without considering the geometric factor, because the structure brings about various functions on a material. Structure can be classified into two categories, external (or global geometry) and internal (or understructure). The former indicates the shape of a component, e.g., films, bars, dots, while the latter does the organization or the pattern in the material, e.g., crystallographic grains, dislocation patterns. It is well known that both structure and size strongly affect fracture behavior.

Before a detailed discussion, we would like to identify our standpoint concerning fracture mechanics on the nanometer scale in terms of material structure. There are mainly two types of issues that we have to consider:

1. Multi-scale investigation focusing on the hierarchical effects of structure:

   The target is a material with the external structure of micrometer to meter size (macromaterial) and the internal one of nanometer size (e.g., nanocrystal). Of course, interaction among different scales of structure (e.g., dislocations/grain boundaries) plays key role in the fracture process.

2. Nano-scale investigation, where the component size itself is in nanometer scale:

   The external structure on the nanometer scale is targeted; e.g., nano-components, such as nano-films, nano-bars, and nano-dots. The internal structures are atomic arrangement and electronic structure in this case. The research aims to elucidate the fracture behavior considering the atomic scale implicitly.

In the context of deformation and fracture on the nanometer scale, these two standpoints share common fundamentals in the mechanics, but remarkably differ in the following points. While the volume decreases in the third power of the reference length with the shrinkage of small components, the surface/interface area does so in the second power. Thus, surface and interface often have critical influence on fracture in nano-components. Moreover, the size difference between "external" and "internal" structure is a crucial factor, although both multi-scale and nano-scale investigations aim to elucidate the fracture process and mechanics under the interaction between phenomena on each scale level. In nano-scale investigation, the scale of stress (or strain) concentration region stemming from the external structure is close to the one of the internal structure (e.g., atomic arrangement). In other words, the internal structure directly affects the fracture mechanism of the whole component. Since this must bring about strong impact on the fracture criterion, one fundamental question that comes up is the applicability of a conventional concept based on the continuum mechanics because of the discrete structure of atomic arrangement.

We can distinctly recognize the philosophical difference in the fracture research between issues 1 and 2 above. The authors, then, would like to emphasize that this book dedicates to the second issue: fracture of nano-components. The discussion on the multi-scale concept (i.e., the first issue) is out of scope because there are already excellent books available on the subject.

This book focuses on the mechanical effect of surface and interface on the characteristics of deformation and fracture.

Because they supply the main sources in the fracture of a bulk, the knowledge and concepts based on conventional continuum mechanics may give us a strong clue for the exploration of the nano-scale world. This is the top-down approach from macro-mechanics to nano-mechanics. On the other hand, considering the discrete internal structure due to the atoms, the definition of stress and strain based on continuum mechanics becomes ambiguous in nano-components. Thus, the lower applicable limit of the continuum mechanics concept is one of interesting issues under the top-down approach. We have to reconsider the definition of fracture and the mechanical parameters on the atomic scale. This requires the bottom-up approach from atomic mechanics (dynamics) to nano-mechanics. Not only the dislocation behavior but also the thermal activation of atomic motion plays an important role in the deformation and fracture processes of nano-components because it has strong size-dependence on deformation (e.g., diffusion creep).

Moreover, we should be careful about the definition of fracture. For example, in an electronic device, a single dislocation can bring about malfunction because of the leak of an electric current. This means that defective nucleation on the atomic scale is the fatal fracture of the whole component. A nano-crack, of course, must be a fatal damage to the component.

Thanks to the recent development of computational power as well as many kinds of computation libraries, numerical simulations based on various techniques provide us information about material behavior. This leads to the reconsideration of material strength from quantum mechanics and reaches the multi-physics concept, where the mechanical aspects interact with the other physical aspects such as electric, electronic, optical, and magnetic. Of course, nano-components possess characteristic behavior due to the external/internal structure as the surface and interface strongly affect multi-physics properties. It suggests one of the future directions of fracture nano-mechanics.

Enormous efforts have been made for the development of meta-materials, comprising numerous elements on the nanometer scale. Since meta-materials have exotic properties, as opposed to ordinary materials, micro-components (films) made up of them have a promising future in the applications of small devices. This suggests the importance of fracture nano-mechanics in pursuing the behavior of aggregated nano-elements in terms of reliability.

1.3 CONVENTIONAL MACRO-MECHANICS

The fundamentals of conventional macro-mechanics are briefly reviewed below.

Deformation

The stress σ_{ij} and the strain ε_{ij} under deformation of material are definitely determined by the following equations: (a) equilibrium equation, (b) compatibility equation, and (c) constitutive equation. As the governing equations of (a) and (b) are derived from Newton's law and from the smoothness (differentiability) of the displacement field, respectively, they are usually valid for ordinary components of continuum media. The characteristic features of material is included in only (c); e.g., Young's modulus in a linear elastic body. Since the unknown in this scheme is the nonlinear plastic behavior, the major effort is put on the investigation of (c) in complex deformation conditions.

Fracture

There is no universal law available even in simple monotonic loading, although the stress- (or strain-) based criterion has been widely adopted for the design of components without a sharp notch or crack. Although the nonlinearity due to plastic deformation and fracture processes makes analysis tougher, the concept usually works well in general. For a cracked body, on the other hand, the stress (or strain) shows a singular field in the vicinity of the crack tip. It is well known that the intensified stress field (the stress intensity factor) governs the crack growth behavior. The mechanics, where the singular stress (or strain) field governs the fracture, is referred to as "fracture mechanics," and this has attained glorious success in the description of cracking behavior. This is, however, a nonlocal criterion, while the stress-based one for the non-cracked (or notched) body is local. The discrepancy brings us complexity and difficulty in the understanding of fracture phenomena; e.g., ceramics have high strength in terms of the local criterion (high fracture stress in a body without a crack), though cracked ones show low toughness (low fracture stress intensity factor in a cracked body). Meanwhile, there is another concept, the energy-based criterion. It has other difficulties, though, such as the energy dissipation due to plasticity and fracture. Thus, we do not have a lucid criterion on the fracture. In fracture analysis, the deformation analysis (stress and/or strain) is ingeniously combined with experimental results on the basis of the above tentative mechanical criteria.

Another difficulty in investigation originates from the fact that the fracture mechanism is not unique. The fracture is not brought about by only monotonic overload. The most popular one in plural mechanisms is "fatigue," where a material fails because of numerous cycles even in a low load range. Time-dependent phenomena such as creep and environmental factor are also very important for the understanding of the fracture mechanism and mechanics. In addition, their interactions, e.g., creep–fatigue interaction and environmental fatigue, make the issue more complex and difficult.

1.4 DEVELOPMENTS IN NANO-MECHANICAL TESTING

With the technological demands increasing for advancement in precision processing, mechanical testing equipment on small materials has been actively developed since the 1990s. The testing method for the downsizing has made remarkable progress, though the fundamental concept is analogical to the one for macro-scale materials. Because of the difficulties in specimen gripping and precise loading, bending type methods, such as four point bend and cantilever bend, have become popular for the nano-scale specimen. The hardness testing method, which can apply a mechanical load at the local site in a material, provides a useful technique in the specific loading. The bending test using the hardness testing tool gives valuable insight into material strength in terms of the mechanics of fracture.

In order to explore fracture law, we need to precisely identify the stress/strain distribution of the specimen in the loading process. In miniature testing, it is difficult to produce the fracture phenomenon under accurate control of mechanical conditions. In nano-scale testing, the focus is on the fabrication and handling technique of the specimen because there are restrictions not only on gripping and loading but also on measurements of displacement and load. As the four point bend method meets the requirement, the strength of interface of a thin film deposited on a substrate has been examined from an early stage of micro-testing. Simplicity in the manufacturing process of the specimen at the scale is also a benefit of the method. However, the four point bend has mm-dimensions except the film thickness. Thus, it might need extra contrivance to apply the method to components with the shape in 3D nano-scale in order to elucidate the effect of the singular field on the nanometer scale on fracture of a real component.

While the electron microscope plays an important role in observation, the scanning probe microscope has a strong impact on the nanometer mechanical testing. A precise control of applied load and displacement of a loading tip by the probe microscope enables us to perform an accurate mechanical testing on the nano-scale thanks to the advancement of the piezo-actuator. By means of the equipment, many investigations of nano-indentation and compression of nano-bars have been developed on the basis of the idea.

The devices for measurement themselves become a kind of MEMS/NEMS. Sharing the benefit of the well-known progress in the fabrication process, we have to design a testing scheme in a total system. While the above is for the specimen produced by the top-down approach, we may apply the technique to mechanical testing by the bottom-up approach in the future.

1.5 DEVELOPMENTS IN NANO-MECHANICAL SIMULATION

Observation of the fracture process and the measurement/evaluation of subtle mechanical variables are extremely tough in nanometer-scale experiments. Numerical analysis offers excellent solutions to this problem.

The finite element method (FEM) and the boundary element method (BEM) provide reliable simulation results of deformation on the basis of continuum mechanics. Since the concept has fundamentally no lower limitation of size, we can get the stress (strain) state even in nano-scale components under the assumption of continuum media. However, there are many difficulties in reality. The constitutive equation of plasticity statistically averages the deformation due to the effect of numerous dislocations. In a nano-component, as an individual dislocation has considerable influence on deformation or the local stress field, the average may lose validity. For modification of continuum mechanics, discrete dislocation dynamics, where the motion and stress field of individual dislocations are numerically pursued, was proposed. Although it is popular in part of multi-scale analysis, it is more attractive as a direct simulation for the analysis of deformation and fracture in nano-components with substructures; e.g., fatigue of nano-metals.

On the other hand, atomic simulations have made great progress in the last decade. Since empirical simple interatomic potentials were popular for simulation until the 1980s, the validity of analysis was questionable in many cases in terms of quantitative evaluation. Taking into account the electrons, the development of semi-empirical potentials under the consideration of many-body interactions pioneers atomic simulation in terms of methodology. It enables us to conduct large-scale simulations and to reproduce material behavior with reasonable precision under comparably light loads. *Ab initio* calculation (first-principles calculation), which numerically solves the Schrödinger equation on the basis of quantum mechanics, has drawn great attention because of its precision and versatility. Not only the enhancement of computer power but also the progress in numerical techniques have greatly contributed to remarkable advancements, including the analysis of mechanical properties of materials. Several teams have given their effort to develop the *ab initio* program for all purposes that has made it easier for researchers in the materials science community to make precise analysis on the atomic scale.

In the atomic analysis at the nano-scale, we have to reconsider the validity and the definition of parameters used in macro-mechanics (stress, strain, elastic constant, etc.) because of the discreteness of atoms.

1.6 STANCE IN THIS BOOK

There has been little investigation on the law of fracture strength for nano-size materials, although many research groups have tackled the deformation and fracture of micro-scale materials in terms of mechanical experiments and analyses. On the other hand, the nano-world possesses enormous potential for more research on unexplored material behavior and for future industrial applications. This mismatch means that the subject discussed in this book is immature and we have so many difficult unknowns. Therefore, this book aims to introduce recent attempts, challenges, and results addressing the mechanics on the nano-scale on the basis of argument and knowledge of macroscopic (conventional) fracture mechanics. We spare many pages on the continuum mechanics concept and the applicability at the nano-scale (top-down) while we introduce the basic concept of materials strength in terms of atomic mechanics (bottom-up).

Since through this book we intend to supply to the readers hints and clues for future investigation, we present the current status of research work rather than fully completed results and accomplished knowledge systematically investigated. For example, we sometimes describe an unpopular experimental method or an unauthorized concept in detail in order to appeal for the necessity of further development. If some of readers get inspiration from the description, it would be our great pleasure. Thus, the primary policy is to contribute for the research evolution on material strength at the nano-scale.

The related subjects are spread diversely, and many excellent books are already published and available. For example, we can readily point out statistical mechanics, fundamental materials science, quantum mechanics, damage mechanics, mechanical testing methodology, numerical simulation technique, and so on. To avoid complexity, we omit the details and restrict the description to only highly relevant knowledge.

Chapter 2

FUNDAMENTALS IN FRACTURE MECHANICS

2.1 FRACTURE MECHANICS

As an introduction to the fracture in the nano-scale, this chapter briefly describes the fundamentals of fracture mechanics, which was developed to describe fractures in macroscopic materials. The chapter covers two main subjects: (a) the singular stress field that brings about the fracture (Sections 2.2–2.5) and (b) applications for a variety of cracking phenomena (Section 2.6).

Conventional fracture mechanics, which deals with the mechanical parameters governing the stress and stress singular field in the vicinity of a crack tip, has greatly increased understanding of crack growth behavior in bulk materials. Extensive experimental evidence shows that the progress of a local fracture (crack growth) is not caused by the stress at a point (crack tip) but by the intensity of the stress (or strain) field. This concept has been developed and expanded to many fracture phenomena and has been used in various industrial sectors to enhance reliability in the design and maintenance of components, such as in the power-generation (e.g., nuclear power plants) and transportation (e.g., aircraft) sectors. As many superb textbooks have been published on fracture mechanics

Fracture Nanomechanics
Edited by Takayuki Kitamura, Hiroyuki Hirakata, Takashi Sumigawa and Takahiro Shimada

www.panstanford.com

for a homogeneous body, we confine the explanation here to the fundamentals closely related to the topics in later chapters.

2.2 FRACTURE MECHANICS IN A HOMOGENEOUS LINEAR-ELASTIC BODY

2.2.1 Singular Stress Field in the Vicinity of the Crack Tip

Figure 2.1 illustrates a crack in a homogeneous linear-elastic body under a remote tensile stress, σ_0. The solid mechanics points out that there appears to be a singular stress field in the vicinity of the crack tip, where the stress magnitude reaches infinity at the tip. The near field is formulated as

$$\begin{pmatrix} \sigma_{xx} \\ \sigma_{yy} \\ \tau_{xy} \end{pmatrix} = \frac{K_{\mathrm{I}}}{\sqrt{2\pi r}} \cos\frac{\theta}{2} \begin{pmatrix} 1-\sin\frac{\theta}{2}\sin\frac{3\theta}{2} \\ 1+\sin\frac{\theta}{2}\sin\frac{3\theta}{2} \\ \sin\frac{\theta}{2}\cos\frac{3\theta}{2} \end{pmatrix} \tag{2.1}$$

where (x, y) and (r, θ) are the Cartesian coordinate and the polar coordinate with the origin at the crack tip, respectively. The important indications are that the field is dependent on the inverse root of the distance from the crack tip, $r^{-1/2}$, and that the intensity of the field is governed by a unique parameter, K_{I}. Thus, K_{I} is the representative fracture mechanics parameter denoted as the (linear-elastic) stress intensity factor.

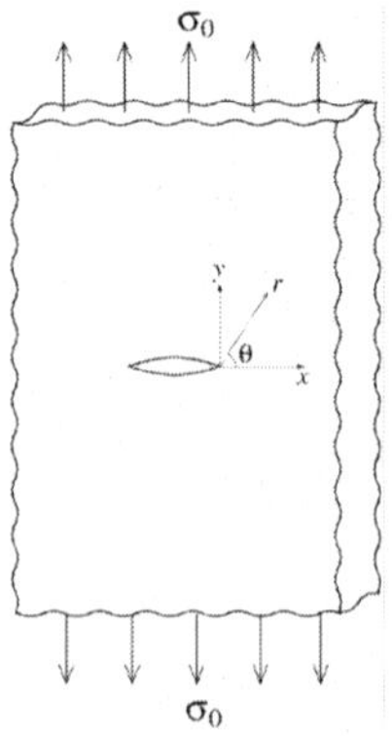

Figure 2.1 Plate with a crack under remote uniform tensile stress.

The crack shown in Fig. 2.1 possesses a symmetric deformation mode on the crack face (opening mode; Fig. 2.2(a)), which is called Mode I. On the other hand, if in-plane or out-of-plane shear is applied on the cracked body, the other deformation modes illustrated in Fig. 2.2 (b) and (c) appear near the crack tip. These deformation modes are denoted as Mode II and Mode III, respectively, and the singular stress fields are given by

$$\begin{pmatrix} \sigma_{xx} \\ \sigma_{yy} \\ \tau_{xy} \end{pmatrix} = \frac{K_{\mathrm{II}}}{\sqrt{2\pi r}} \begin{pmatrix} -\sin\frac{\theta}{2}\left(2+\cos\frac{\theta}{2}\cos\frac{3\theta}{2}\right) \\ \sin\frac{\theta}{2}\cos\frac{\theta}{2}\cos\frac{3\theta}{2} \\ \cos\frac{\theta}{2}\left(1-\sin\frac{\theta}{2}\sin\frac{3\theta}{2}\right) \end{pmatrix} \tag{2.2}$$

and

$$\begin{pmatrix} \tau_{xz} \\ \tau_{yz} \end{pmatrix} = \frac{K_{\mathrm{III}}}{\sqrt{2\pi r}} \begin{pmatrix} -\sin\frac{\theta}{2} \\ \cos\frac{\theta}{2} \end{pmatrix} \tag{2.3}$$

respectively. The stress fields possess the singularity of inverse root r, and their intensities are represented by the stress intensity factors K_{II} and K_{III}, respectively. The stress fields in the vicinity of a crack tip under general loading in a homogeneous linear-elastic body (mixed mode) are shown by the superposition of Eqs. (2.1) to (2.3). In other words, the singular field can be clearly decomposed into Mode I, Mode II, and Mode III, and K_{I}, K_{II}, and K_{III} govern the respective intensities.

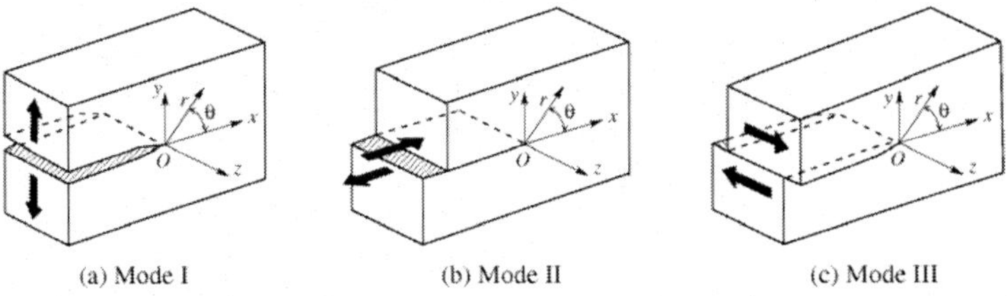

Figure 2.2 Deformation modes at the crack tip. (a) The opening mode; Mode I; (b) the in-plane shear mode, Mode II; and (c) the out-of-plane shear mode, Mode III.

It is important that this can be generalized for an arbitrary condition, which means that the stress (strain) field in the vicinity of the tip of any straight crack in any component under any boundary condition is always uniquely represented by the combination of Eqs. (2.1) to (2.3).

The fracture mechanics concept is based on the idea that cracking behavior is governed by stress (strain) intensity. Thus, once we find the dependence of the cracking behavior on the intensity (K_I, K_{II}, K_{III}) through experimentation, we can apply the relationship to a crack in a real component.

The intensities in the simplest case, where a wide plate with a crack at the center under remote tension, σ_0, as shown in Fig. 2.1, is given as

$$(K_I, K_{II}, K_{III}) = (\sigma_0\sqrt{\pi a}, 0, 0) \tag{2.4}$$

In general, K is dependent on the boundary condition; e.g., the component shape. Then, the intensity of Mode I is expressed in the form

$$K_I = F\sigma\sqrt{\pi a} \tag{2.5}$$

where F is a function of the crack shape, the loading condition, and the component shape. Using handbooks (e.g., *Stress Intensity Factors Handbook*, The Society of Materials Science, Japan [JSMS], vols. 1 and 2 published by Pergamon Press [1987], vol. 3 published by Pergamon Press and JSMS [1993], and vols. 4 and 5 published by JSMS [2001]) that collect and list the stress intensity factors in various conditions, we can easily evaluate the magnitudes from tables without performing stress/strain analysis such as the finite element method (FEM) and boundary element method (BEM).

Since the plastic deformation appears in a high-stress area, the actual stress field in the vicinity of the crack tip departs from those expressed in Eqs. (2.1) to (2.3). However, when the deviation is small (e.g., small-scale yielding (SSY), which is localized yielding near the tip), the equations approximately represent the field, except the local plastic area.

In the same manner, infinite stress never actually appears in a real component; the K field discussed above approximates the real distribution affected by the process zone (the localized fracture-proceeding area near the crack tip).

Numerous experimental studies have proved that the cracking behavior is controlled by the fracture mechanics parameters, K_I, K_{II}, and K_{III}, under the elasticity-dominated condition.

2.2.2 Energy Release Rate

During the crack propagation by an incremental length, Δa, the strain energy stored is released from the body. Here, the potential energy released, $\Delta\Pi$, which is equal to the work done by the force needed to close the tip of the extended crack, as shown in Fig. 2.3 (b), is in the function of Δa. Then, the energy release rate, G, is defined as

$$G = -\lim_{\Delta a \to 0} \frac{\Delta\Pi}{\Delta a} = -\frac{\partial\Pi}{\partial a} \tag{2.6}$$

In the case of Mode I, G correlates with the stress intensity factor, K_I, as

$$G = \frac{K_I^2}{E'} \tag{2.7}$$

$$E' = E \quad \text{in the plane stress} \tag{2.8}$$
$$E' = E/(1-\nu^2) \quad \text{in the plane strain}$$

Under the mixed mode, G is given as the superimposed ones in each mode

$$G = \frac{K_I^2 + K_{II}^2}{E'} + \frac{K_{III}^2}{2\mu} \tag{2.9}$$

where μ is the shear modulus.

On the other hand, excess energy is required to make a new surface in the crack growth. The generalized Griffith theory [Griffith, 1920] assumes that the excess energy is supplied by the released potential energy. Thus, the criterion for crack growth is given by

$$G = 2\gamma \tag{2.10}$$

where γ is the surface energy. The criterion can be represented by the stress intensity factors as the correlation of Eq. (2.9). Since G is applicable to any mode of cracking, the concept unifies the criteria of stress intensity factors. Equation (2.10), however, applies when there is no energy dissipation. In general, the energy dissipates in the yielding zone and the process zone, where the irreversible deformation and/or complex fracture processes go on, so that the energy balance is not exact and the criterion, Eq. (2.10), loses physical meaning.

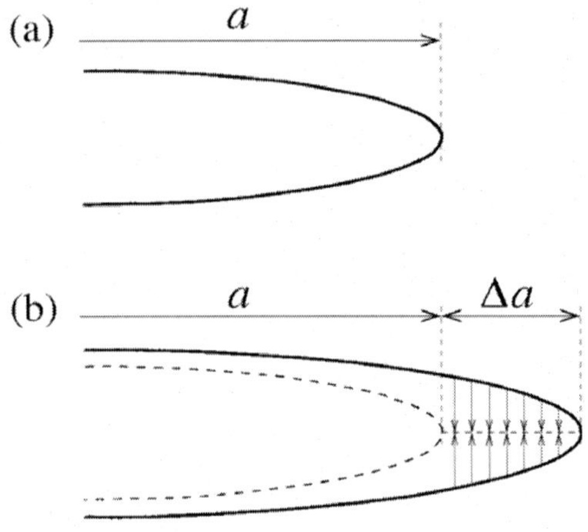

Figure 2.3 Crack extension and energy released (a) before the extension, and (b) after the extension.

2.2.3 *J*-Integral

Rice [1968a] proposed a fracture mechanics parameter, the *J*-integral (*J*) which is defined by

$$J \equiv \int_{\Gamma} \left(W \mathrm{d}y - T_i \frac{\partial u_i}{\partial x} \mathrm{d}s \right) \tag{2.11}$$

$$W = \int_0^{\varepsilon_{ij}} \sigma_{ij} \mathrm{d}\varepsilon_{ij} \tag{2.12}$$

where Γ is the integral path from the lower crack face to the upper crack face surrounding the crack tip, as illustrated in Fig. 2.4; W is the strain energy density; σ_{ij} is the stress tensor; ε_{ij} is the strain tensor; T_i is the component of traction; and u_i is the component of displacement. The magnitude of the *J*-integral is independent of the integral path, which indicates that the information near the crack tip can be evaluated using the mechanical quantities far from the tip.

It has been proven that the *J*-integral is identical to the energy release rate in the linear-elastic body [Rice, 1968b]:

$$J = G \tag{2.13}$$

Of course, the J-integral is correlated with the stress intensity factors, as shown in Eqs. (2.7) to (2.9).

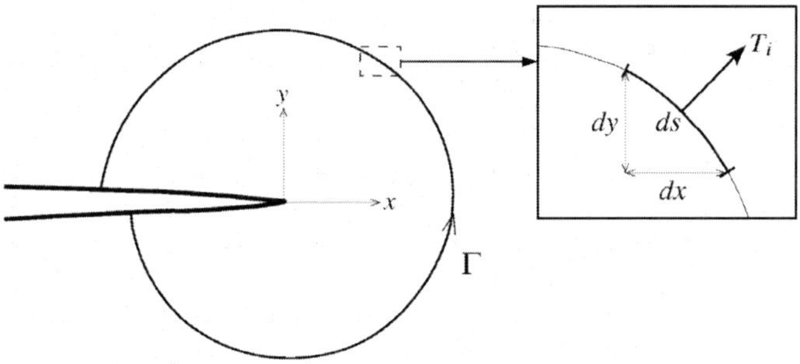

Figure 2.4 The integral path of the J-integral surrounding the crack tip.

2.3 FRACTURE MECHANICS IN A HOMOGENEOUS ELASTIC-PLASTIC BODY

2.3.1 Stress Singularity in Power Law Plasticity

Although fracture mechanics in an elastic-plastic body, where nonlinearity and irreversibility arise in the stress-strain relation, has not been fully established, numerical analysis provides us an insight into the complexity of the issues. Thus, the discussion in the elastic-plastic stress field has steadily progressed, albeit under various restrictions. Although current understanding of the issues is incomplete, it is worth noting that many experiments complement the analysis and underscore the essential validity of the concept.

If the stress at any point in a body does not decrease, the plasticity under the total strain theory brings the same stress and strain fields as that in a nonlinear-elastic body with the same σ to ε relation (the constitutive law). Hutchinson [1968] and Rice and Rosengren [1968] analyzed the singular stress and strain fields of an elastic-plastic body assuming the power law constitutive relation (Fig. 2.5),

$$\varepsilon = \sigma/E + B\sigma^n \tag{2.14}$$

where B and n are the material constants, representing the plastic characteristics. They derived the following stress fields, denoted as HRR singularity, in the vicinity of the crack tip in the Modes I and II under a nondecreasing load on the basis of J_2 deformation theory:

$$\sigma_{ij} = \frac{K_{\mathrm{p}\sigma}}{r^{1/(n+1)}} \tilde{\sigma}_{ij}(\theta) \tag{2.15}$$

$$\varepsilon_{ij} = \frac{K_{\mathrm{p}\varepsilon}}{r^{n/(n+1)}} \tilde{\varepsilon}_{ij}(\theta) \tag{2.16}$$

Here, $K_{\mathrm{p}\sigma}$ and $K_{\mathrm{p}\varepsilon}$ are the plastic (nonlinear elastic) stress intensity factor and the plastic (nonlinear elastic) strain intensity factor; $\tilde{\sigma}_{ij}(\theta)$ and $\tilde{\varepsilon}_{ij}(\theta)$ are the functions of θ. Although the dependence of singularity on r, $r^{-1/(n+1)}$, is different from that in a linear-elastic body, $r^{-1/2}$, they are similar in that the intensity of singular stress (strain) is governed by the magnitude of single coefficient, the stress (strain) intensity factor.

In a nonlinear elastic body with a crack, the J-integral, J, defined by Eq. (2.11), keeps the path independency. Moreover, it has been proved that J is equivalent to the energy release rate, as defined in Eq. (2.6). This suggests the applicability of J-integral on the elastic-plastic body. In reality, there exists the relation

$$J = I_n B\left(K_{\mathrm{p}\sigma}\right)^{n+1} \text{ and } J = \frac{I_n}{B^{1/n}}\left(K_{\mathrm{p}\varepsilon}\right)^{n+1/n} \tag{2.17}$$

Then, substituting Eq. (2.17) into Eqs. (2.15) and (2.16), we have the singular fields,

$$\sigma_{ij} = \left(\frac{J}{I_n B r}\right)^{1/(n+1)} \tilde{\sigma}_{ij}(\theta) \tag{2.18}$$

$$\varepsilon_{ij} = B\left(\frac{J}{I_n B r}\right)^{n/(n+1)} \tilde{\varepsilon}_{ij}(\theta) \tag{2.19}$$

Here, I_n is the function of n. The singular field is characterized by the fracture mechanics parameter, J.

This is why the cracking behavior in a homogeneous elastic-plastic body under a nondecreasing load is usually analyzed on the basis of the HRR field and the J-integral, although the energy balance loses its physical meaning. Numerous experimental results under elastic-plastic conditions have supported the validity of the J approach.

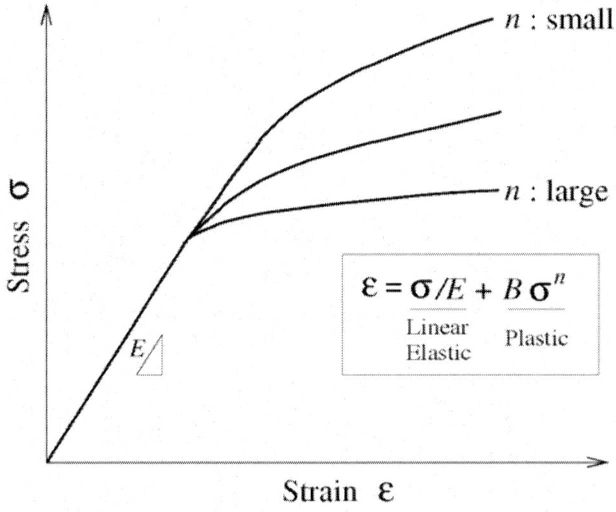

Figure 2.5 Schematic view of typical stress-strain relation (Eq. 2.14) in an elastic-plastic body. The exponent, *n*, represents work-hardening behavior.

2.3.2 Stress Singularity in Power Law Creep

It is well known that strain increases under a constant load at an elevated temperature as time elapses, which is called creep. In electronic devices, creep deformation takes place in metals with a low melting temperature and in polymers used at moderate temperatures. Because creep rate is time dependent and nonlinearly dependent on the stress in general, a complex behavior appears in the stress field in the vicinity of the crack tip.

2.3.2.1 Static creep

Figure 2.6 shows a creep curve under a constant tensile load (static creep) at a constant temperature. After transient creep (primary creep), where the creep rate decreases as time elapses, the rate becomes constant. Called steady state creep or secondary creep, this actually occupies the greater part of creep life. In the following discussion, we show the stress field under steady state creep, which follows the power law constitutive relation,

$$\dot{\varepsilon} = B_c \sigma^{n_c} \tag{2.20}$$

where $\dot{\varepsilon}$ is the creep rate, and B_c and n_c are the material constants.

Equation (2.20) resembles the plastic law formulated in Eq. (2.14). Converting the strain and the displacement in the plastic (nonlinear-elastic) analyses into the strain rate and the displacement rate in the creep one is called the Hoff analogy [Hoff, 1954].

Applying the Hoff analogy to a cracked body, we have the analogical HRR stress field near the crack tip in the creep. The singular fields of the stress and the strain rate, then, have the forms

$$\sigma_{ij} = \left(\frac{J^*}{I_n B_c r} \right)^{1/(n_c+1)} \tilde{\sigma}_{ij}(\theta) \tag{2.21}$$

$$\dot{\varepsilon}_{ij} = B_c \left(\frac{J^*}{I_n B_c r} \right)^{n_c/(n_c+1)} \tilde{\varepsilon}_{ij}(\theta) \tag{2.22}$$

[Ohji et al., 1976; Landes and Begley, 1976], where J^* is the creep J-integral defined by

$$J^* \equiv \int_\Gamma \left(W^* \mathrm{d}y - T_i \frac{\partial \dot{u}_i}{\partial x} \mathrm{d}s \right) \tag{2.23}$$

$$W^* = \int_0^{\dot{\varepsilon}_{ij}} \sigma_{ij} \mathrm{d}\dot{\varepsilon}_{ij} \tag{2.24}$$

Here, W^* is the strain energy rate density, $\dot{u}_i$ is the displacement rate, and $\dot{\varepsilon}_{ij}$ is the strain rate. J^* is, of course, the path-independent integral in a fully crept body.

2.3.2.2 Transition from SSC to LSC

At the loading instant ($t = 0$), we have the elastic stress field (no creep effect) in the vicinity of the crack tip, Eq. (2.1), because creep is a time-dependent phenomenon. On the other hand, Eq. (2.21) shows the steady state under creep-dominated conditions after enough time has elapsed. This signifies that the stress field changes from an elastic condition (Eq. (2.1)) to a creep condition (Eq. (2.21)) after loading, as schematically illustrated in Fig. 2.7. As faster creep takes place in the high stress region, the stress change begins from the crack tip and expands over time. In other words, the creep-dominated zone appears from the crack tip (Fig. 2.7(b)) and grows (Fig. 2.7(c)) until it covers the greater part of the ligament (Fig. 2.7(d)). This is the transition from small-scale creep (SSC) to large-scale creep (LSC). Assuming that the stress field in the creep-dominated zone follows Eq. (2.21), J^* in the transition (Fig. 2.8) is given by [Ohji, K. et al., 1980; Riedel and Rice, 1980]

$$J^* = \frac{K^2}{(n+1)Et} \tag{2.25}$$

It is important to note that J^* is time-dependent in the transition. Here, where the integral path of J^* is inside the creep region, the magnitude is independent of the integral path. J^*, however, has path-dependence outside the creep region.

If the body is surrounded by a hard material, the creep is constrained and the creep zone cannot expand. In that case, the stress field remains in an SSC condition and the cracking behavior is governed by K, much like an elastic-plastic body under small-scale yielding [Uematsu et al., 1995].

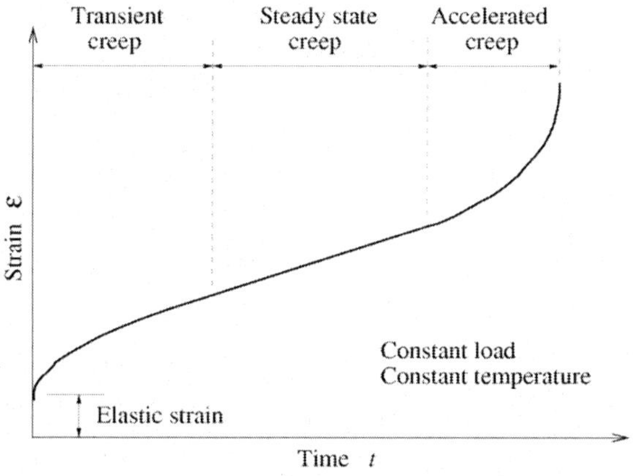

Figure 2.6 The creep curve consists of three stages: transient, steady, and accelerated.

Primary creep appears in every cycle under fatigue loading at an elevated temperature (creep-fatigue condition) and brings about complex time dependence on the stress field. However, there are few reliable studies on the stress singularity of a cracked body because the constitutive relation under the primary creep is intricate.

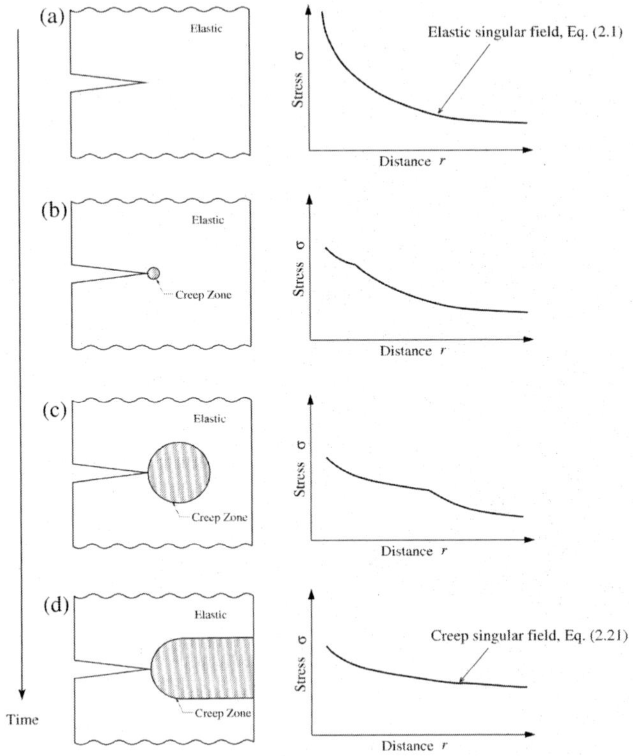

Figure 2.7 Transition from small-scale creep to large-scale creep.

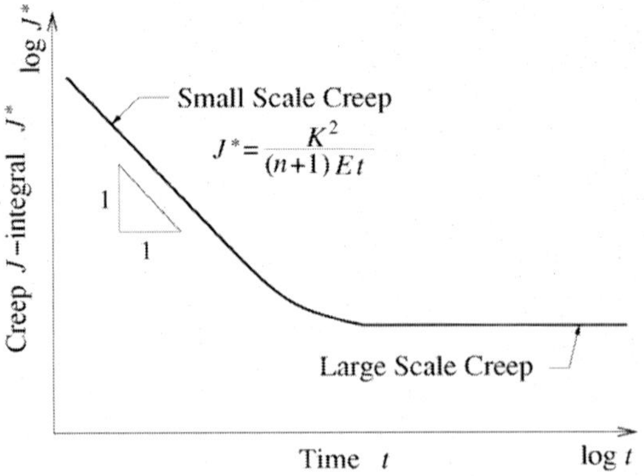

Figure 2.8 Change in the creep J-integral during the transition from SSC to LSC.

2.4 FRACTURE MECHANICS ON AN INTERFACE CRACK

2.4.1 Stress Singularity along the Interface of an Elastic Bi-material

This section discusses the stress field near the tip of an interface crack in a bi-material, as illustrated in Fig. 2.9. The elastic constants—such as the Young's modulus, E; shear modulus, μ; and Poisson's ratio, ν of each material (materials 1 or 2)—are distinguished by the suffix. The following conditions are assumed for the connection of materials at the interface:

$$u_{x1} = u_{x2} \text{ and } u_{y1} = u_{y2} \tag{2.26}$$

$$T_{x1} + T_{x2} = 0 \text{ and } T_{y1} + T_{y2} = 0 \tag{2.27}$$

These indicate the continuity of displacement and the equilibrium of traction, where (u_{xj}, u_{yj}) and (T_{xj}, T_{yj}) are the displacement and traction of a material j (j = 1 or 2). In other words, the interface has perfect bonding and no sliding is allowed along it.

For the crack, the stress field in the vicinity of the tip along the interface ($\theta = 0$) [Erdogan, 1963] is given by

$$\sigma_{yy} + \mathbf{i}\tau_{xy} = \frac{K_1 + \mathbf{i}K_2}{\sqrt{2\pi r}}\left(\frac{r}{l}\right)^{\mathrm{i}\varepsilon} \tag{2.28}$$

$$\varepsilon = \frac{1}{2\pi}\ln\left(\frac{1-\beta}{1+\beta}\right) \tag{2.29}$$

where β is one of the Dunders parameters [Dunders, 1967] shown in Section 2.5.1, Eq. (2.37). The dependence of θ is given in the original paper [Sun and Jih, 1987]. Here, $\mathbf{i}$ is the imaginary unit, l is a reference length, and $K_1 + \mathbf{i}K_2$ is the complex stress intensity factor. Equations (2.28) signify that the field has the singularity of $r^{-1/2}$ and the intensity is uniquely represented by the pair of stress intensity factors K_1 and K_2. Although this seems similar to the formula for the crack in a homogeneous body, there is great difference between them. In Eq. (2.28) $(r/l)^{\mathrm{i}\varepsilon}$ indicates the oscillation, where the singular field vibrates between the two envelopes with the singularity of $r^{-1/2}$, shown in Fig. 2.10 by a solid line. Moreover, there is an overlap between the crack faces behind the crack tip.

Modes I and II cannot be separated in Eq. (2.28), which means that the singular stress fields cannot be expressed by the superposition of pure modes in general. In the other words, we always have a mixed mode singularity under the load. Thus, we must be careful that the stress intensity (K_1, K_2) is different from (K_I, K_{II}) as discussed in Section 2.2.1, and that (K_1, K_2) does not represent the intensity of the opening mode and the in-plane shear mode. Although we can have an arbitrary length for l that expresses oscillate singularity in a nondimensional form, the crack length is usually used for l for convenience.

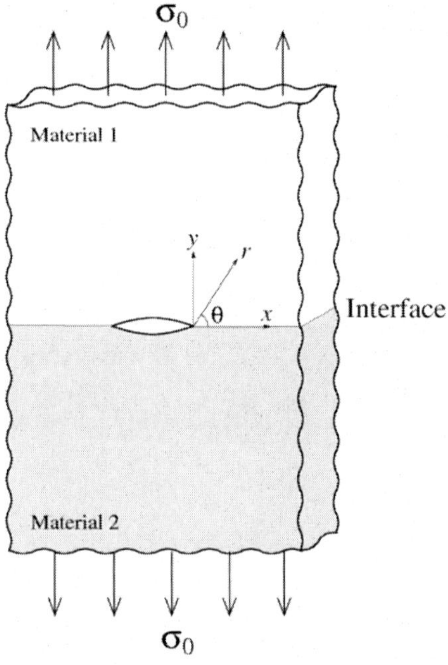

Figure 2.9 Bi-material plate with an interface crack under uniform remote tensile stress.

The effect of oscillation in the stress distribution, in general, is confined to a very tiny zone in the immediate neighborhood of the crack tip. In a real crack, because the plastic strain (or process zone) dominates there, the oscillation may not appear. Thus, using the concept of small-scale yielding, we can discuss the cracking behavior on the basis of the complex stress intensity factor (K_1, K_2), which represents the stress intensity of the envelop stress field. We can understand the disappearance of the unphysical behavior, such as the overlap of crack faces near the tip, with the same reasoning.

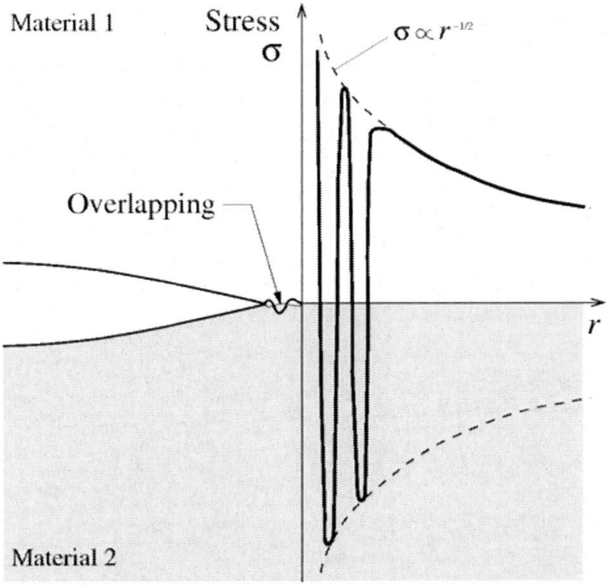

Figure 2.10 Oscillated stress singular field near an interface crack tip.

2.4.2 Energy Release Rate

Like the cracking in a homogeneous body shown in Fig. 2.3, the strain energy is released by the growth, as illustrated in Fig. 2.11 (a) and (b). The released energy, $\Delta\Pi$, is the function of crack growth length, Δa, and the energy release rate, G, is defined for the interface crack by

$$G = -\lim_{\Delta a \to 0} \frac{\Delta\Pi}{\Delta a} = -\frac{\partial\Pi}{\partial a} \tag{2.6}$$

The energy release rate is correlated with the complex stress intensity factor by [Malyshev and Salganik, 1965],

$$G = \frac{1}{16\cosh^2(\varepsilon\pi)}\left\{\frac{\chi_1+1}{\mu_1}+\frac{\chi_2+1}{\mu_2}\right\}\left(K_1^2+K_2^2\right) \tag{2.30}$$

where χ_i is shown in Section 2.5.1, Eq. (2.38). Because it is not easy to identify experimentally the growth criterion of interface crack on the basis of (K_1, K_2) due to the mixed mode, G is often used as a unified governing parameter. Although the procedure is convenient, it should be noted that it cannot be strictly applied to the crack growth with the energy dispersion, e.g., plastic deformation.

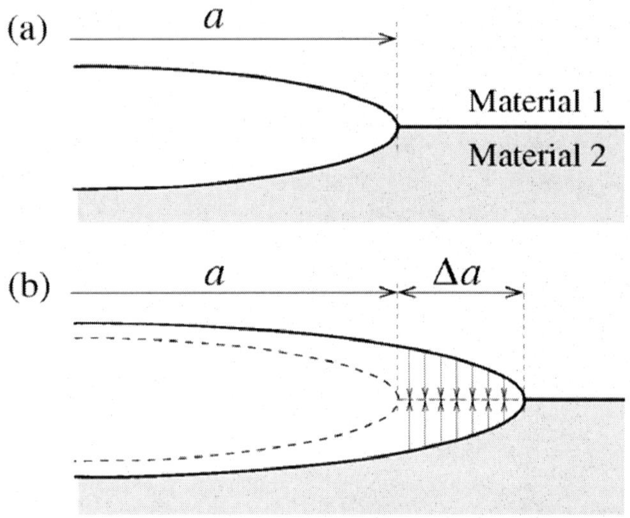

Figure 2.11 Extension of interface crack and energy released (a) before the extension, and (b) after the extension.

2.4.3 Stress Singularity in Power Law Plastic Bi-material

The singular stress field in the vicinity of the interface crack tip under elastic-plastic conditions is too intricate to give us a general analytical solution. Current research mainly depends on numerical computer simulation.

Shih and Asaro [1988] conducted a pioneering analysis for bi-material following the power laws (see Eq. (2.14)) using a finite element method and found a similar singular field without the oscillation found in the HRR field. Later, detailed investigations have revealed that the stress field in the immediate vicinity of the interface crack tip obeys

$$\sigma_{ij} = \frac{K_{\mathrm{p}\sigma_{(\mathrm{interface})}}}{r^{1/(n_2+1)}}\tilde{\sigma}_{ij}(\theta) \tag{2.31}$$

$$\varepsilon_{ij} = \frac{K_{\mathrm{p}\varepsilon_{(\mathrm{interface})}}}{r^{n_2/(n_2+1)}}\tilde{\varepsilon}_{ij}(\theta) \tag{2.32}$$

under $n_1 < n_2$, where n_1 and n_2 denote n of materials 1 and 2, respectively. This means that the softer behavior of materials governs the singular characteristics. Since the intensity of the singular stress field can be represented by the coefficient, $K_{p\sigma(\text{interface})}$, conventional fracture mechanics can be applied to cracking. For a discussion on the validity of path-independent integrals, including the J-integral, see Gao and Lou [1990].

For an interface crack in a creep bi-material, a field similar to the HRR is reported, although few experiments have been conducted [Biner, 1997]. Of course, the transition from SSC to LSC brings about a change in the singular stress field.

2.5 STRESS SINGULARITY IN THE VICINITY OF INTERFACE EDGE

2.5.1 Elastic Bi-material

Figure 2.12 shows schematic views of a bi-material with a junction between an interface and a free surface, which is called the interface edge. The deformation mismatch concentrates the strain near the interface edge (free-edge effect). A singular stress field often appears in a bi-material with a flat surface (Fig. 2.12(a)), while it does not in the surface of a homogeneous body. In general, the stress singularity in the following form is often found near an interface edge

$$\sigma_{ij} \propto r^{-\lambda} = r^{p-1} \tag{2.33}$$

where r is the distance from the interface edge. Bogy [1968 and 1971] found that p is given by

$$C_A\beta^2 + 2C\alpha\beta + C_C\alpha^2 - 2C_D\beta - 2C_E\alpha + C_F = 0 \tag{2.34}$$

$$C_A = 4H(p, \theta_1)H(p, \theta_2)$$

$$C_B = 2p^2 \sin\theta_1\, H(p, \theta_2) + 2p^2 \sin\theta_2\, \mathrm{H}(p, \theta_1)$$

$$C_C = 4p^2(p^2 - 1)\sin^2\theta_1 + \sin^2\theta_2 + H(p, \theta_1 - \theta_2) \tag{2.35}$$

$$C_D = 2p^2\, \{\sin^2\theta_1 \sin^2(p\theta_2) - \sin^2\theta_2 \sin^2(p\theta_1)\}$$

$$C_E = -\,C_D + H(p, \theta_2) - H(p, \theta_1)$$

$$C_{\mathrm{F}} = H(p, \theta_1 + \theta_2)$$
$$H(p, \theta) = \sin^2(p\theta) - p^2 \sin^2\theta \tag{2.36}$$

where α and β are the Dunders parameters [Dunders, 1967] which have the following correlation with the elastic coefficients of materials 1 and 2, μ_1 and μ_2 (shear modulus) and ν_1 and ν_2 (Poisson's ratio), respectively.

$$\alpha = \frac{\mu_1(\chi_2+1)-\mu_2(\chi_1+1)}{\mu_1(\chi_2+1)+\mu_2(\chi_1+1)} \tag{2.37}$$

$$\beta = \frac{\mu_1(\chi_2-1)-\mu_2(\chi_1-1)}{\mu_1(\chi_2+1)+\mu_2(\chi_1+1)}$$

$$\chi_i = 3 - 4\nu_i \text{ for plane strain}$$
$$= \frac{3-\nu_i}{1+\nu_i} \text{ for plane stress} \tag{2.38}$$

The characteristic of the stress field, namely the power λ in Eq. (2.33), is dependent not only on the combination of material but also on the shape of the edge (the angles of materials 1 and 2 near the interface edge, φ_1 and φ_2). $\lambda = 1 - p$, which is evaluated by Eq. (2.34), is not equal to 1/2 in general. The singularity appears in $0 < p < 1$, while it does not in $1 < p$. It should be noted that Eq. (2.34) sometimes has two solutions, which means the combined singularity appears in the vicinity of the interface edge. Moreover, p sometimes is an imaginary number, which indicates oscillated singularity (see the stress singularity for the interface crack, Fig. 2.10).

Figure 2.13 shows the magnitude of λ for the combinations of (α, β) and $(\varphi_1, \varphi_2) = (\pi/2, \pi/2)$. Because λ is dependent not only on (α, β) but also on (φ_1, φ_2), the stress intensity factor in Eq. (2.33), K, can represent only the intensity for a specific shape of the interface edge between a specific material combination. This signifies that the critical intensity, K_c, must be determined for each (α, β) and (φ_1, φ_2) according to conventional fracture mechanics.

In terms of industrial applications of this concept, it is inconvenient for engineers to determine the toughness of components (critical intensity) for each (α, β) and (φ_1, φ_2). The criterion, moreover, cannot be used to quantitatively compare interface strength because the dimension of stress intensity factor is dependent on (α,β) and (φ_1,φ_2). If we could find an energy-based criterion, it might be used to avoid the dependence on the shape of interface and might be universally used to determine the crack initiation of a particular interface. Moreover, because it represents the toughness of the interface, a comparison among different interfaces by their magnitudes would become possible. However, the energy release rate has not been successfully defined for the initiation of an interface crack from the edge in general.

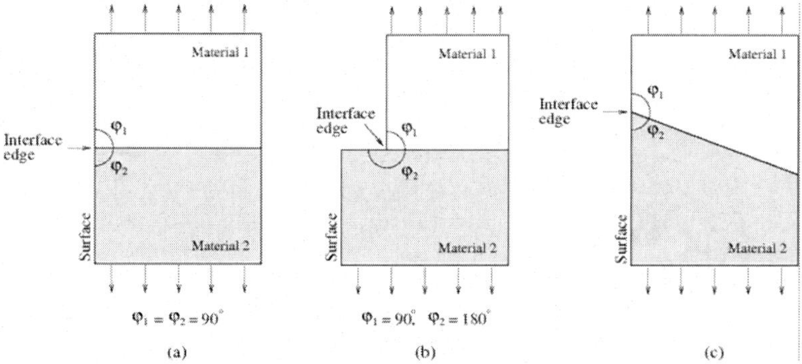

Figure 2.12 Bi-material plate with the free interface edge, which is the junction of the interface face and the surface.

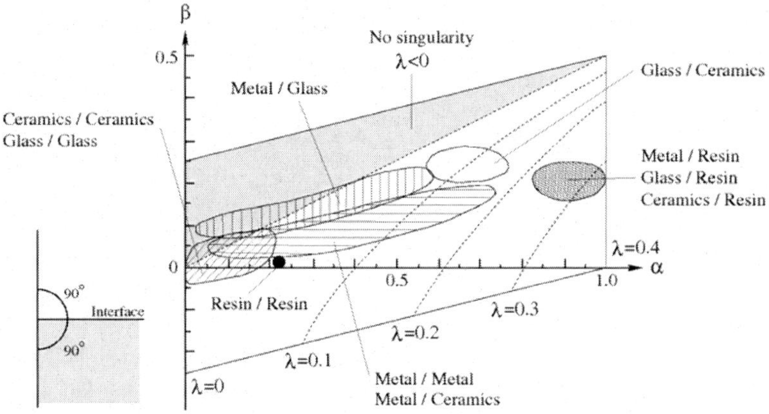

Figure 2.13 Dependence of stress singularity on the elastic property of bi-material in the interface of $(\varphi_1, \varphi_2) = (\pi/2, \pi/2)$ (Fig 2.12(a)).

2.5.2 Stress Singularity in Power Law Plastic Bi-material

Like the analysis of the singular stress field near the interface crack in an elastic/plastic (or plastic/plastic) bi-material, there are many unknowns about the stress field near an interface edge. Several case studies [Wang, 1990; Reedy, 1993] have been conducted using numerical simulations, and equations such as

$$\sigma_{ij} \propto r^{-\lambda} = r^{p-1} \tag{2.39}$$

were derived. A systematic product, however, has not been extracted.

As discussed in Section 2.4.3, because material 1 is more compliant than material 2 when n_2 is significantly smaller than n_1, material 1 dominates singular behavior near the interface edge. Thus, an analysis [Xu et al., 2002] was carried out on a bi-material with the interface edge assuming that material 2 was a rigid solid. The singularity, λ, is dependent on the hardening exponent, n_1, and the bonded angle of material 1, φ_1; the magnitudes are given in Fig. 2.14. This elucidates that the smaller n_1 is, the stronger the singularity is, and that λ converges $\frac{1}{n+1}$ at large φ_1.

This can be applied in a creep/creep (or creep/elastic) bi-material when the creep-dominated area expands far enough from the interface edge (LSC), and the result of elastic/plastic bi-material can be used. The transient behavior from SSC to LSC has been numerically discussed [Kitamura et al., 2007 and 2008], with the point made that the stress intensity factor decreases in proportion of $t^{1/(n+1)}$, similar to Eq. (2.25).

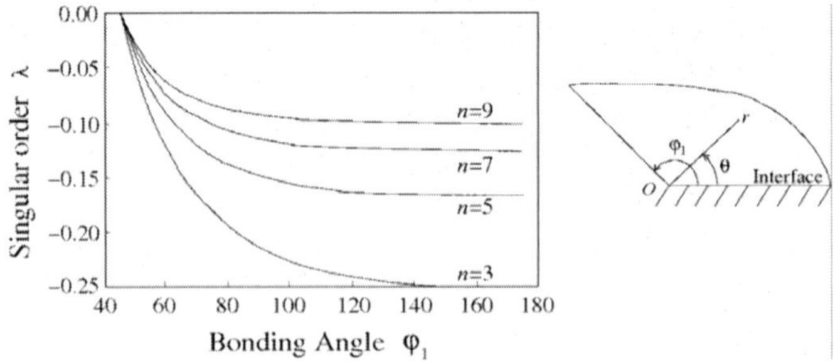

Figure 2.14 Dependence of singular order on bonding angle in elastic-plastic conditions [Xu et al., 2002].

2.6 APPLICATION OF FRACTURE MECHANICS

2.6.1 Fracture Toughness

Consider a material with a crack subjected to a monotonically increasing load. The (linear-elastic) stress intensity factor *K* (or the *J*-integral *J* if it is elastic-plastic), which characterizes the intensity of the singular stress field near the crack tip, increases with an increase of the applied load. This is the driving force for crack propagation, and the crack begins to propagate when the magnitude of *K* (*J* or *G*) reaches the material's inherent resistance to crack propagation, namely, the fracture toughness, K_C (J_C or G_C). In particular, toughness is designated as K_{IC} under Mode I loading. Thus, the initiation criterion of crack propagation is expressed as the following equation:

$$K \geq K_C \text{ (or } J \geq J_C,\ G \geq G_C) \tag{2.40}$$

Figure 2.15(a) is a schematic diagram of the relationship between the crack driving force and the crack extension resistance. The solid lines in the figure designate the crack extension resistance *R* (*R*-curve), while the dashed lines represent the crack driving force *K*. In brittle materials, *K* increases as the applied load increases, and unstable crack propagation takes place at point A without stable crack growth from the initial crack length a_0. In ductile materials, the crack tip is blunted up to point A and begins to stably propagate at point A. Because the resistance, *R*, is higher than the crack driving force *K* after point A, additional *K* is required to extend the crack further. Unstable crack propagation then occurs at point B. The magnitude of *K* (*J* or *G*) at the initiation of the unstable crack propagation for brittle materials and that of the stable crack propagation for ductile materials are defined as the fracture toughness K_C (J_C or G_C). If *K* is a decreasing function of *a*, for example, in the double cantilever specimen under a constant displacement, the crack is arrested when *K* becomes lower than *R*, as shown at point C in Fig. 2.15(b). The resistance to crack propagation along a bi-material interface, i.e., interface fracture toughness, can be similarly evaluated. The stress field at an interface crack between thin film and substrate is likely to be the plane strain condition, and constraint by the substrates often causes brittle delamination.

2.6.2 Subcritical Crack Growth

A crack can gradually propagate under a much lower load than the fracture toughness when fatigue, creep, and a corrosive environment are factors. Therefore, the application of fracture mechanics to these phenomena has a great importance.

2.6.2.1 Fatigue

In fatigue loading, cyclic plastic deformation in the vicinity of the crack tip causes a crack to grow. Under SSY conditions, the fatigue crack growth rate, da/dN, is characterized by the stress intensity factor range ΔK $(= K_{max} - K_{min})$ where K_{max} and K_{min} are the maximum and minimum stress intensity factors.

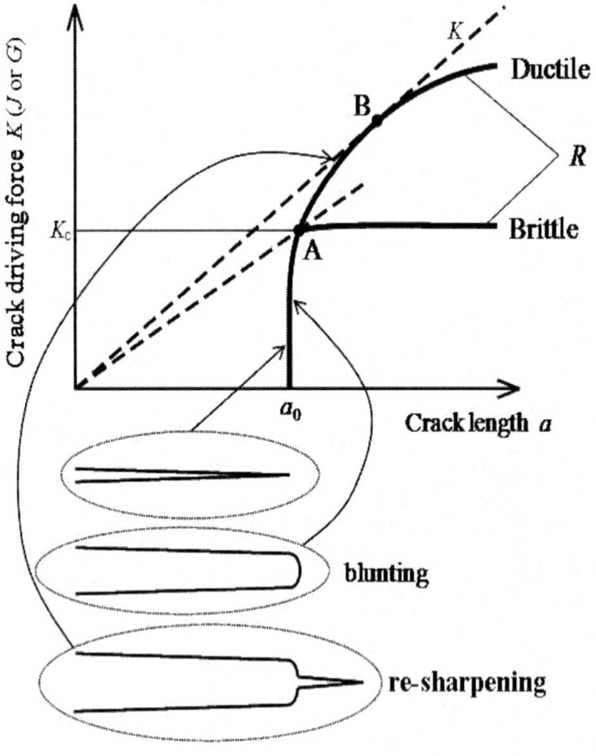

(a) Load control

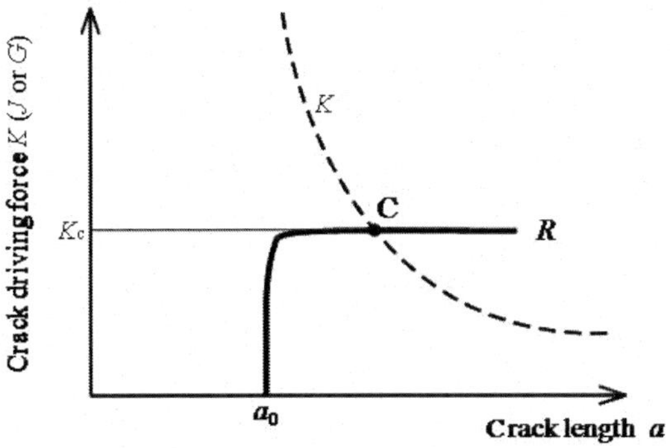

(b) Displacement control

Figure 2.15 Crack extension force and resistance (*R*-curve).

The relationship between da/dN and ΔK is typically represented by the sigmoid curve shown in Fig. 2.16 and divided into three regions: the threshold region (A), the stable growth region (B), and the critical growth region (C). In region A, there is generally a lower limit of fatigue crack propagation, and da/dN rapidly decreases near the threshold stress intensity factor, ΔK_{th}. In region B, the $da/dN - \Delta K$ relation is expressed by the Paris law [Paris and Erdogan, 1963]:

$$da/dN = C\Delta K^m \tag{2.41}$$

In the Paris law, C and m are the material constants characterizing fatigue crack growth behavior, and they are dependent on the stress ratio $R = K_{min}/K_{max}$. In this region, crack-tip blunting and re-sharpening take place in each cycle, and therefore the striations are typically observed on the fracture surfaces. In region C, da/dN rapidly increases when K_{max} reaches the fatigue fracture toughness K_{fC}, which is close to the fracture toughness K_C.

Because a fatigue crack has a plastic zone ahead of the crack tip, the growth leaves tensile plastic strain in the region near the crack face behind the tip. Therefore, the crack near the tip closes in the unloading process even if the applied load is tensile; namely, the plasticity-induced crack closure. There are other crack closure mechanisms as well, such as oxide-induced crack closure, roughness-induced crack closure, and fluid-induced crack closure. In those cases, only the stress intensity factor range where the crack opens contributes to fatigue crack growth, and the range is designated as the effective stress intensity factor range ΔK_{eff} (= $K_{max} - K_{op}$), where K_{op} is the stress intensity factor to open the whole crack closure region [Elber, 1971]. The *da/dN* is reasonably correlated with ΔK_{eff}, and the relationship is not dependent on the stress ratio *R*. In this case, the Paris law is rewritten as

$$da/dN = C' \Delta K_{eff}^{m'} \tag{2.42}$$

where *C'* and *m'* are the material constants.

Under the LSY condition, which is often classified as low-cycle fatigue, *da/dN* is characterized by the *J*-integral range (cyclic *J*-integral) ΔJ [Dowling and Begley, 1976], which is the *J*-integral in the tension-going half cycle. When the crack tip closes, only the process above the crack opening load is taken into account. In this case, the crack growth rate is expressed as

$$da/dN = C_J \Delta J^{m_J} \tag{2.43}$$

where C_J and m_J are material constants. Since ΔJ is expressed by $\Delta K_{eff}^2/E'$ under the SSY condition, ΔJ can be applied to a wider range of materials than ΔK.

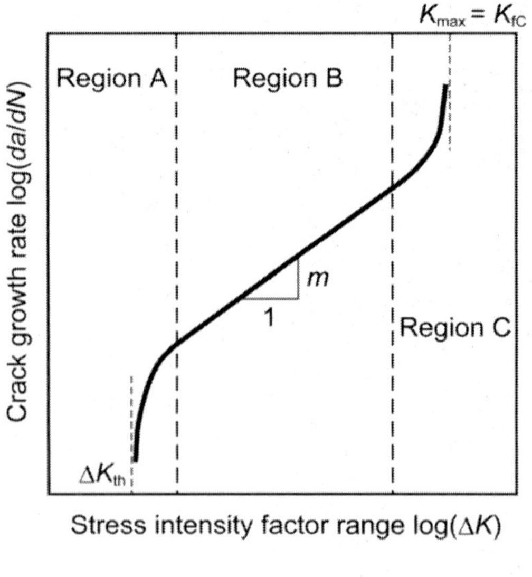

Figure 2.16 Schematic diagram showing the typical relationship between the fatigue crack growth rate and the stress intensity factor range.

2.6.2.2 Creep

At an elevated temperature, a crack time-dependently grows due to creep even under a low load. The creep crack growth rate, *da/dt,* is dominated by the singular stress (or strain rate) field of Eq. (2.21) (or Eq. (2.22)) under the large-scale creep (LSC) condition. Therefore, the *da/dt* is characterized by the creep *J*-integral J^* (Eqs. (2.23) and (2.24)), which represents the intensity of the singular field [Taira et al., 1979a]. In this case, the relationship is expressed by the following law:

$$da/dt = C_c(J^*)^{m_c} \tag{2.44}$$

Here, C_c and m_c are the material constants characterizing the creep crack growth behavior, and for many materials m_c is almost unity. The effect of the transition from the small-scale creep (SSC) to LSC can be evaluated using the creep *J*-integral J^* in the transition state. J^* in the transition state is calculated by the same equation as Eqs. (2.23) and (2.24), where the integration path Γ is set within the creep-dominated region (see Fig. 2.7) near the crack tip.

The creep crack grows under the SSC condition without transition in special cases, e.g., delamination cracks in fiber-reinforced plastics (FRP), where the creep deformation near the crack tip is strictly constrained by the presence of fibers [Uematsu et al., 1995]. In that case, the creep crack growth is dominated by the elastic stress intensity factor *K*, and the following relationship is found between *da/dt* and *K*.

$$da/dt = C_{SSC}(K)^{m_{SSC}} \tag{2.45}$$

Here, C_{SSC} and m_{SSC} are the material constants. This may take place in a micro-device where a creeping material is surrounded by hard materials.

In high-temperature fatigue, the combination of creep and fatigue brings about intricate cracking behavior [Taira et al., 1979b], namely, creep-fatigue interaction.

2.6.2.3 Environment-assisted crack growth

When a material is subjected to tensile stress in a corrosive environment, a crack is initiated and propagates under the critical load, which is known as stress-corrosion cracking (SCC). The two main mechanisms of SCC are active-path corrosion (APC-SCC) and hydrogen embrittlement (HE-SCC). In APC-SCC, a crack grows through localized corrosion because of the anodic dissolution in the stress-concentrated region near the crack tip. In HE-SCC, hydrogen diffusing into the material concentrates on the region near stress-concentrated sites such as notches, inclusions, and precipitates, and this leads to the cracking.

The crack growth rate da/dt due to SCC is governed by environmental and mechanical factors. The mechanical factor is characterized by the stress intensity factor K. Figure 2.17 shows the schematic relationship between da/dt and K, where the curve is divided into three regions, A, B, and C. In region A, the da/dt monotonically increases with an increase in K, and the lower limit for stress corrosion cracking is designated as K_{ISCC} under Mode I loading. In region B, the plateau region, da/dt is independent of K because crack growth is controlled by rate-limiting processes such as chemical reaction and mass transport. In region C, the crack growth rate accelerates and K converges to K_{IX}, which is close to the fracture toughness K_C in an inert environment. Because da/dt in region B is generally fast, the threshold K_{ISCC} is the most important property in SCC.

Fatigue crack growth resistance decreases in corrosive environments; this is known as corrosion fatigue [Rolfe and Barsom, 1977; Minoshima and Komai, 1988]. Cracks easily propagate by the synergistic effect of cyclic loading and the corrosive environment.

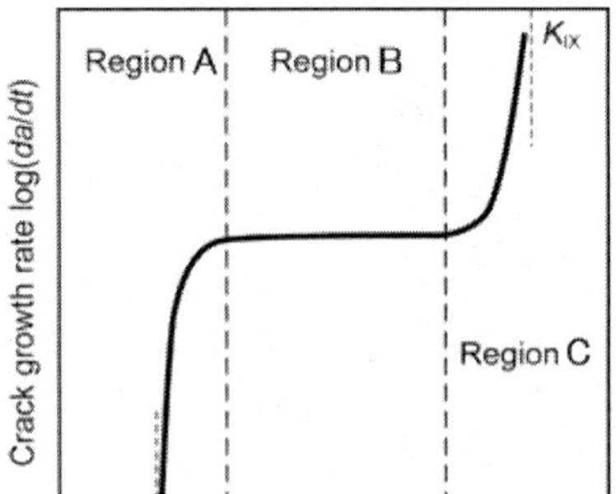

Figure 2.17 Schematic diagram showing the typical relationship between SCC crack growth rate and stress intensity factor.

References

Biner, S. B. (1997) A numerical analysis of time dependent stress fields ahead of stationary cracks at creep regime. Part I: Interface cracks in bi-materials, *Eng. Fract. Mech.*, 56, pp. 513–529; and Part II: Interface cracks in layered materials, *Eng. Fract. Mech.*, 56, pp. 657–674.

Bogy, D. B. (1968) Edge bonded dissimilar orthogonal elastic wedges under normal and shear loading, *J. Appl. Mech.*, 35, pp. 146–154.

Bogy, D. B. (1971) Two edge bonded elastic wedges of different materials and wedge angle under surface traction, *J. Appl. Mech.*, 38, pp. 377–386.

Elber, W. (1971) The significance of fatigue crack closure, *ASTM STP 486*, pp. 230–242.

Erdogan, F. (1963) Stress distribution in a nonhomogeneous elastic plane with cracks, *J. Appl. Mech.*, 30, pp. 232–238.

Dunders, J. (1967) Effect of elastic constants on stress in a composite under plane deformation, *J. Compo. Mater.*, 1, p. 310.

Dowling, N. E., and Begley, J. A. (1976) Fatigue crack growth during gross plasticity and the *J*-integral, *ASTM STP 590*, pp. 82–103.

Gao, Y., and Lou Z. (1990) Mixed mode interface crack in a pure power-hardening bimaterial, *Int. J. Fract.*, 43, pp. 241–256.

Griffith, A. A. (1920) The phenomena of rupture and flows in solids, *Philosophical Transactions of the Royal Society of London*, A221, pp. 163–198.

Hoff, N. J. (1954) Approximate analysis of structures in the presence of moderately large creep deformation, *Quart. Appl. Math.*, pp. 49–55.

Hutchinson, J. W. (1968) Singular behaviour at the end of a tensile crack in a hardening material, *J. Mech. Phys. Solids*, 16, pp. 13–31; and (1968) Plastic stress and strain fields at a crack tip, *J. Mech. Phys. Solids*, 16, pp. 337–347.

Kitamura, T., Ngampangpis, K., and Hirakata, H. (2007) Stress field near interface edge of elastic-creep bi-material, *Eng. Fract. Mech.*, 74, pp. 1637–1648.

Kitamura, T., Ngampangpis, K., and Hirakata, H. (2008) Increase of stress intensity near interface edge of elastic-creep bi-material under a sustained load, *Eng. Fract. Mech.*, 75, pp. 1285–1293.

Landes, J. D., and Begley, J. A. (1976) A fracture mechanics approach to creep crack growth, *ASTM STP 590*, pp. 128–148.

Malyshev, B., and Salganik, R. (1965) The strength of adhesive joints using the theory of cracks, *Int. J. Fract. Mech.*, 1, pp. 114–119.

Minoshima, K., and Komai, K. (1988) Stress corrosion cracking of high-strength steels under dynamic loading conditions, in *Localized Corrosion (Current Japanese Materials Research),* vol. 4 (ed. Hine, F., Komai, K., and Yamakawa, K.), Elsevier Applied Science, London, p. 123.

Ohji, K., Ogura, K., and Kubo, S. (1976) Analysis on growth of mode III crack in creep, *Trans. Jap. Soc. Mech. Eng.*, 42, pp. 350–358.

Ohji, K., Ogura, K., and Kubo, S. (1980) Stress field and modified *J*-integral near a crack tip under condition of confined creep deformation, *J. Soc. Mater. Sci., Japan*, 29, pp. 465–471.

Paris, P. C., and Erdogan, F. (1963) A critical analysis of crack propagation laws, *Trans. ASME, Ser. D,* 85, pp. 528–534.

Reedy, E. D., Jr. (1993) Free-edge stress intensity factor for a bonded ductile layer subjected to shear, *J. Appl. Mech.*, 60, pp. 715–720.

Rice, J. R. (1968a) A path independent integral and the approximate analysis of strain concentration by notches and cracks, *Trans. ASME, J. Appl. Mech.*, 35, pp. 379–386.

Rice, J. R. (1968b) Mathematical analysis in the mechanics of fracture, in *Fracture* (ed. Libowitz, H.), Academic Press, New York, vol. 2, pp. 191–311.

Rice, J. R., and Rosengren G. F. (1968) Plane strain deformation near a crack tip in a power-law hardening material, *J. Mech. Phys. Solids*, 16, pp. 1–12.

Riedel, H., and Rice, J. R. (1980) Tensile cracks in creeping solids, *ASTM STP 700*, pp. 112–130.

Rolfe, S. T., and Barsom, J. M. (1977) *Fracture and Fatigue Control in Structures*, Prentice-Hall, Upper Saddle River, NJ.

Shih, C. F., and Asaro, R. J. (1988) Elastic-plastic analysis of cracks on biomaterial interfaces. Part I: Small-scale yielding, *J. Appl. Mech.*, 55, pp. 299–316; and (1989) Part II: Structure of small-scale yielding fields, *J. Appl. Mech.*, 56, pp. 763–779.

Sun, C. T., and Jih, C. J. (1987) On strain energy release rate for interfacial cracks in bi-material media, *Eng. Fract. Mech.*, 28, pp. 13–21.

Taira, S., Ohtani, R., and Kitamura, T. (1979a) Application of *J*-integral to high-temperature crack propagation. Part I: Creep crack propagation, *Trans. ASME, J. of Eng. Mater. Tech.*, 101, pp. 154–161.

Taira, S., Ohtani, R., and Komatsu, T. (1979b) Application of *J*-integral to high-temperature crack propagation. Part II: Fatigue crack propagation, *Trans. ASME, J. Eng. Mater. Tech.*, 101, pp. 154–161.

Uematsu, Y., Kitamura, T., and Ohtani, R. (1995) Delamination behavior of a carbon-fiber-reinforced thermoplastic polymer at high temperatures, *Comp. Sci. Tech.*, 53, pp. 333–341.

Wang, T. C. (1990) Elastic-plastic asymptotic fields for cracks on biomaterial interfaces, *Eng. Fract. Mech.*, 37, pp. 527–538.

Xu, J., Fu, L., and Mutoh, Y. (2002) A method for determining elastic-plastic stress singularity at the interface edge of bonded power law hardening materials, *JSME Int. J.*, Ser. A, 45, pp. 177–183.

Chapter 3

ELASTO-PLASTIC DEFORMATION AND FRACTURE OF THIN FILM

3.1 MECHANICAL PROPERTY IN NANO-SCALE

In order to investigate the fracture criterion in terms of the nano-mechanics, we must analyze precisely the stress field near the failure sites such as crack tips and interface corners. The key to this is understanding deformation—elasticity, plasticity, creep, and so on—on a nanometer scale. That is, the deformation of nano-scale materials is quite different from that of their bulk counterparts, so an experimental evaluation is required.

In this chapter, after we briefly introduce the typical deposition techniques of thin films, we review the experimental methods used to evaluate the mechanical properties of thin films. While our focus is on deformation properties, we also briefly discuss fracture strength, because of the similarity in experimental methods.

3.2 FABRICATION TECHNIQUES OF THIN FILM

The deposition methods of nano- or submicron-thick films are roughly classified into two types: gas-phase processing and liquid-phase processing. Each encompasses several techniques, as shown

Fracture Nanomechanics
Edited by Takayuki Kitamura, Hiroyuki Hirakata, Takashi Sumigawa and Takahiro Shimada

www.panstanford.com

in Fig. 3.1. Several important ones are briefly explained below. For more detail, see references [e.g., Freund and Suresh, 2003; Eckertova, 1986].

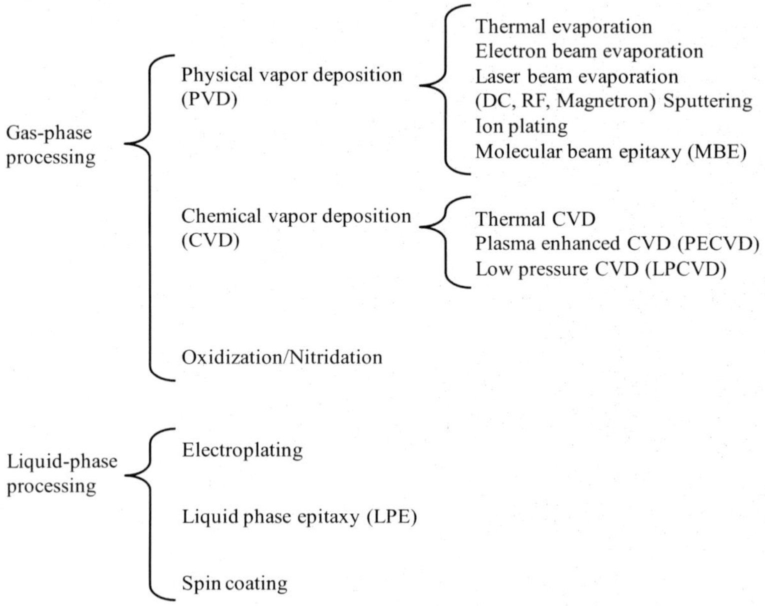

Figure 3.1 Thin film deposition methods.

1. Physical vapor deposition (PVD)

A physical process such as evaporation or sputtering without a chemical reaction generates a vapor that is deposited on a substrate. In the case of evaporation, the atoms are pulled out from the source material using thermal energy by resistive heating (thermal evaporation), electron beam (electron beam evaporation), laser beam (laser beam evaporation), or similar methods. After the atoms travel a distance in a vacuum chamber, they diffuse and condense on the substrate surface. In the sputtering technique, the source vapor is supplied through ionic impingement on the target material. Ar gas in a chamber is ionized by an electrical direct current (DC) or radio-frequency (RF) voltage to form a plasma, and the Ar^+ ions, accelerated toward the target by electrical field, sputter out the atoms from the source material by bombarding them. These atoms then travel through the gas phase and condense on the substrate.

The PVD method can produce thin films made of various metals and ceramics. The microstructure in the formed films is influenced by many factors, such as the base gas pressure, the working pressure, the substrate temperature, and the deposition rate.

2. Chemical vapor deposition (CVD)

The chemical vapor deposition method produces a film by causing a chemical reaction on the substrate (film) surface. The method enables the deposition of elemental and compound semiconductors, metal alloys, and amorphous or crystalline compounds of different stoichiometry. Low-pressure CVD (LPCVD) and plasma-enhanced CVD (PECVD) are variants of the CVD method used for particular purposes.

3. Oxidation (nitridation)

During oxidation or nitridation, a thin film is formed by heat treatment at high temperature in an oxidation (nitrogen) atmosphere. The advantage of this method is that a very pure, thin film without defect (high-quality film) can be obtained without the use of any special device. Large internal stress is sometimes introduced in the film because of the volume expansion induced by the chemical reaction.

4. Electroplating

Electroplating involves coating a conductive surface with a metal through electrochemical reactions facilitated by an applied electrical potential. The surface, which is immersed in a solution of metal salts, must be conductive and cathodic in the electrical circuit. When an electrical current passes through the solution, the positive ions of the source metal are attracted to the cathode surface, forming a thin film.

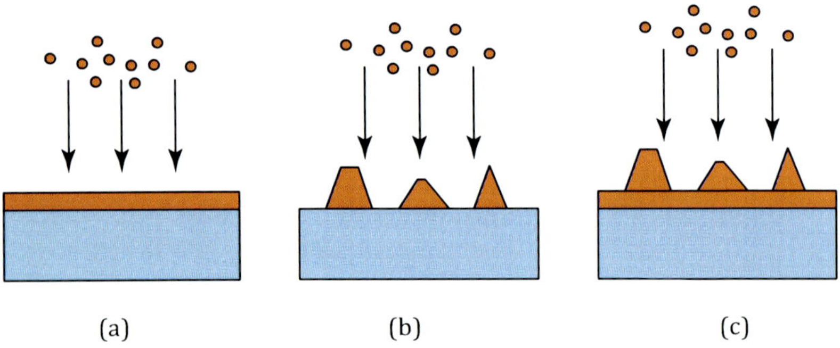

Figure 3.2 Growth mechanisms of thin films. (a) Frank-van der Merwe mode. (b) Volmer-Weber mode. (c) Stranski-Krastanov mode.

Film growth mechanisms can be classified into three modes as illustrated in Fig. 3.2: (a) the Frank-van der Merwe mode, (b) the Volmer-Weber mode, and (c) the Stranski-Krastanov mode, depending on the interaction between the atoms and between the atoms and the substrate. In most cases, the evolution during vapor deposition follows the Volmer-Weber mode. After crystalline islands nucleate from the diffusing atoms on the substrate surface, they grow individually until they meet with other islands. The coalesced islands form plural grains, bringing about a polycrystalline thin film consisting of columnar-orientated crystals. The microstructure of thin films can be modified through a post-deposition process, e.g., annealing.

Since the mechanical properties of thin film strongly depend on the microstructures—such as the grain size and shape, the crystallographic structure, the crystallographic orientation, the crystal imperfection (crystal or amorphous), and the density of defects (dislocations, impurities, and voids)—we must pay careful attention to the deposition technique and the detailed conditions of its application.

3.3 MECHANICAL TESTING

Table 3.1 Methods for evaluating the mechanical properties of thin films

Method	Specimen	Merits	Drawbacks
Tensile test	Free-standing film	Direct measurement of stress and strain	Fabrication and handling of free-standing film
Bending test	Free-standing film	Use of conventional nanoindenter or AFM	Fabrication of free-standing film
Bulge test	Free-standing film	Easy sample handling	Fabrication of free-standing film
Bilayer tensile test	Film on substrate	Easy sample handling	Consideration of substrate effect
Bilayer bending test	Film on substrate	Use of conventional nanoindenter or AFM	Consideration of substrate effect
Indentation	Film on substrate	Easy sample preparation Use of conventional nanoindenter or AFM	Analysis of complex contact deformation Consideration of substrate effect
Thermal stress method	Film on substrate	Easy sample preparation	Change in temperature during the test Consideration of substrate effect
Surface acoustic wave method	Film on substrate	Easy sample preparation	Only elastic property
Vibration lead method	Free-standing film/Film on substrate	Easy sample handling	Only elastic property
Bending of cantilever	Small component	Use of conventional nanoindenter or AFM	Sample preparation by FIB or lithography
Compression of bar	Small component	Use of conventional nanoindenter or AFM	Sample preparation by FIB or lithography

We usually use the tensile test to evaluate the mechanical properties of a bulk material. Because the stress and strain are uniform in the specimen and the tested volume is large, we can readily examine various mechanical properties related to deformation and fracture strength. The accurate evaluation of thin films, however, is not so easy because of the difficulties in the fabrication and handling of a free-standing film specimen. Moreover, the technique of precise light-load applications is awkward, including chucking of specimen, actuator, and load/displacement sensors. In recent years, these problems have been overcome by MEMS technology, and tensile tests for submicron-thick films have been reported [e.g., Haque and Saif, 2002a].

Evaluation methods for a film coupled with a substrate have also been developed [e.g., Florande and Nix, 2005], where the mechanical property of substrate is well known. The film properties are estimated by analysis based on the mechanical response of a film/substrate specimen. Although specimen preparation and experimental setup are easy when compared with testing free-standing film, estimation errors may result from modeling inaccuracy in the analysis.

Several methods for examining the mechanical properties of smaller components, e.g., nano-cantilever and nano-bar, have been proposed [e.g., Uchic et al., 2004; Greer and Nix, 2006; Hirakata et al., 2007].

In the rest of this section, we briefly review mechanical testing methods for thin films and small components. Table 3.1 lists the typical methods and their merits and drawbacks.

3.3.1 Tensile Test

A tensile load is directly applied to a free-standing film as shown in Fig. 3.3, and the elongation of the film is measured by an extensometer. Obtaining the stress-strain relationship is straightforward because the fields in the film are uniform and unidirectional. This can be applied not only when monotonic tension is being evaluated but also when evaluating creep at high temperature, deformation in certain environmental conditions, and fatigue under cyclic loading.

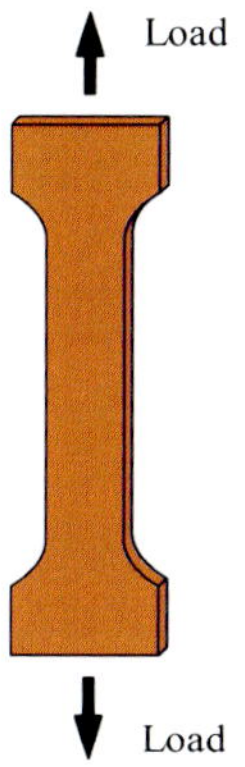

Figure 3.3 Schematic illustration of a tensile test for thin films.

Although conventional mechanical loading apparatuses with high precision are available, the difficulties lie in specimen fabrication and load alignment and gripping. The popular process of fabrication is dissolving a sacrificial layer between the film and the substrate by wet or dry etching. To avoid testing difficulties, MEMS-based techniques have been proposed for the production of loading apparatus and displacement sensors [e.g., Haque and Saif, 2002a]. Methods using a conventional nanoindenter or AFM have also been proposed [e.g., Espinosa et al., 2003a].

3.3.2 Bending Test

As shown in Fig. 3.4, an indenter applies a bending load to a free-standing film, of which one or both edges are fixed to the substrate [Namazu et al., 2002; Zhang et al., 2000]. The free-standing specimens are fabricated using a micromachining lithography technique. Because the films are fixed on the substrate in each type, handling is easy in comparison with handling in the tensile test. In addition, conventional indenters or AFM are available to apply the load. Stress analysis is required to estimate the mechanical properties of the film because the stress field is not uniform.

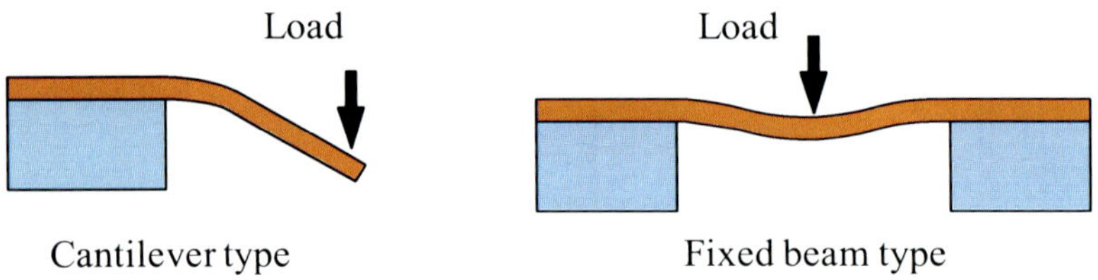

Figure 3.4 Schematic illustrations of bending tests. (a) Cantilever type. (b) Fixed beam type.

3.3.3 Bulge Test

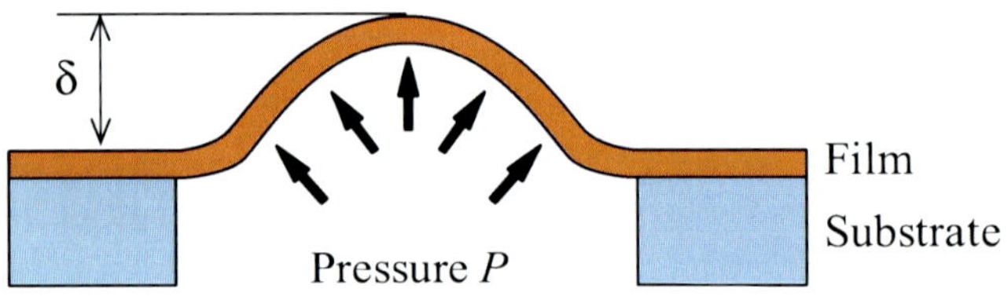

Figure 3.5 Schematic illustration of a bulge test.

Free-standing film with an open window in the substrate is prepared using micromachining and etching techniques. The film is deflected by pressure through the hole in the substrate, as shown in Fig. 3.5 [Tabata et al., 1989; Vlassak and Nix, 1992]. The mechanical properties are evaluated by the pressure-deflection relation. The benefit is the simplicity of specimen handling while the drawback is the stress (strain) distribution in the film. A detailed analysis of the deflection behavior is required to estimate the material's properties.

Assuming spherical film geometry, a simple evaluation is applicable to the stress in the film. Measuring the radius, a, and height of swelling, δ, the stress, σ, and the strain, ε, of thin film are given in the elastic deformation as follows:

$$\sigma = \frac{a^2 P}{4 t_f \delta} \tag{3.1}$$

$$\varepsilon = \frac{2\delta^2}{3a^2} \tag{3.2}$$

where P and t_f are the pressure and the film thickness, respectively. For a more precise estimation, numerical calculation such as the FEM and BEM is needed to analyze the behavior.

In a conventional bulge test with a circular window, the film is in a plane equibiaxial stress state. By using the rectangle window, we can evaluate the mechanical response under uniform plane strain deformation (plane strain bulge test) [Xiang et al., 2005].

3.3.4 Bilayer Tensile and Bending Tests

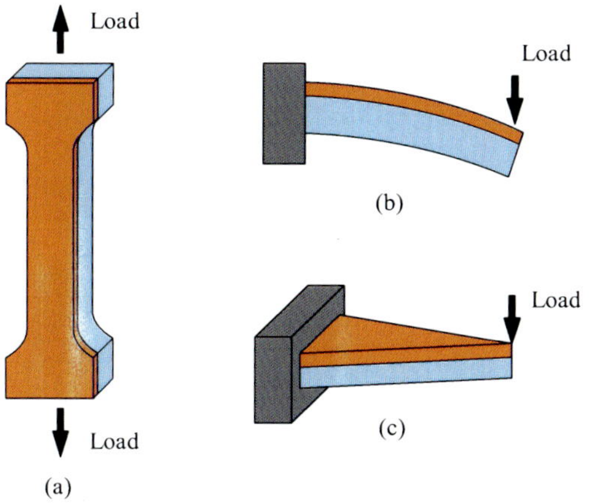

Figure 3.6 Schematic illustrations of bilayer tensile and bending tests.

As shown in Fig. 3.6, a thin film is deposited on a compliant substrate, and a tensile or bending load is applied to the bilayer specimen [Yu and Spaepen, 2004; Griffin et al., 1992; Florando and Nix, 2005]. The handling and the experiment are easier than those for free-standing films. In addition, the methods can be applied to ultra-thin or nano-films for which free-standing experiments are extremely difficult. However, the effect of the compliant substrate has to be considered to precisely evaluate the material properties of the thin film.

In bilayer bending for a triangular specimen as shown in Fig. 3.6c, the stress is constant in the longitudinal direction due to uniform moment, i.e., the uniform strength beam [Florando and Nix, 2005].

3.3.5 Indentation Test

Mechanical properties are evaluated from the load-displacement response in the indentation by a rigid indenter, as shown in Fig. 3.7. This popular method is widely used because of the simplicity and ease of specimen preparation and loading. A three-sided pyramid diamond tip is favored over a four-sided Vickers indenter because it is easy to make a sharp apex. A Berkovich tip with a face angle of 65.3° or effective cone angle of 70.3° is often used because the tip has the same projected area-to-depth ratio as the Vickers indenter. A hemisphere tip is sometimes used in order to apply the Hertzian contact theory.

To be exact, it is necessary to analyze the complex stress/strain state near the contact region when estimating material's properties. However, simplified estimation has been proposed. For example, using the gradient of the load-displacement curve under the unloading S shown in Fig. 3.6b, the reduced modulus,

$$E^* = \left(\frac{1-\nu^2}{E} + \frac{1-\nu_i^2}{E_i} \right)^{-1} \tag{3.3}$$

is estimated [Oliver and Pharr, 1992] by

$$E^* = \frac{1}{\beta'} \frac{\sqrt{\pi}}{2} \frac{S}{\sqrt{A}} \tag{3.4}$$

$$A = 24.5 h_c^2 \tag{3.5}$$

$$h_c = h_{max} - \varepsilon' \frac{P_{max}}{S} \tag{3.6}$$

where E, ν, E_i, and ν_i are the Young's modulus and the Poisson's ratio of the thin film, and the indenter, A is the projected contact area, P_{max}

and h_{max} are the maximum load and displacement, and h_c is the real penetration depth as shown in Fig. 3.7c. The Young's modulus of the film can be estimated if the Poisson's ratio is known. The correction factor β' depends on the Poisson's ratio of the indented material and the tip angle ($\beta' = 1.067$ for $\nu = 0.3$ for the Berkovich indenter), and the factor ε' is 0.75 for the Berkovich indenter [Hay et al., 1999].

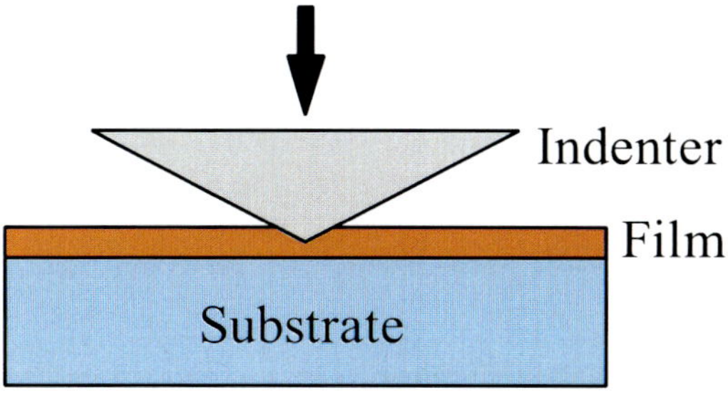

(a)

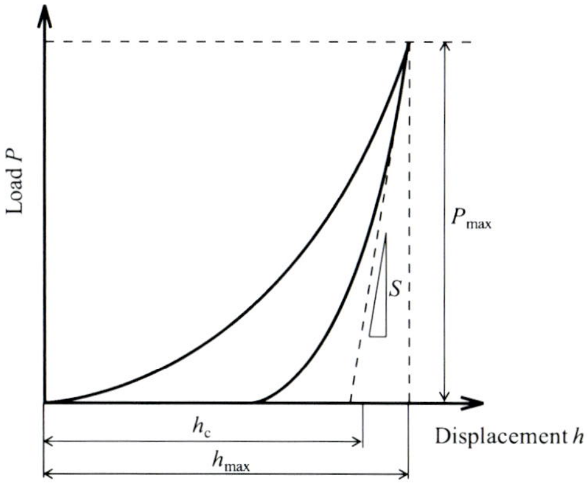

(b)

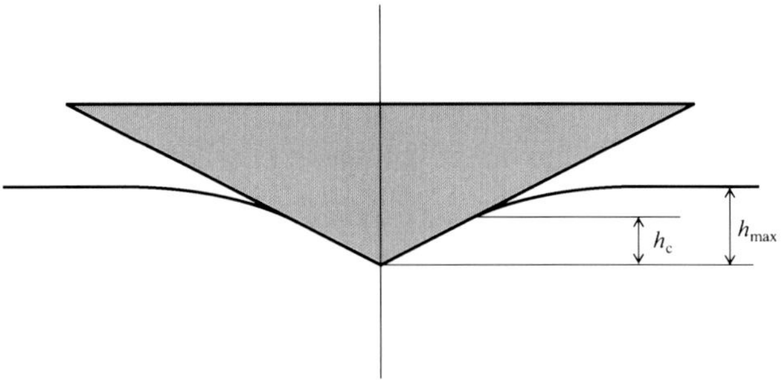

(c)

Figure 3.7 Schematic illustrations of an indentation test.

In evaluating the mechanical properties of nano-film, we should pay attention to the influence of the substrate [Tunvisut et al., 2002; Tsui et al., 1999; Saha and Nix, 2002; Chen and Vlassak, 2001]. We recommend that the penetration depth should not exceed 10% of film thickness to avoid that influence. Other difficulties of accurate evaluation include that the result is very sensitive to the tip shape, the loading angle, the surface roughness, and the complex contact state.

Although this method has been extended to the evaluation of plastic properties, it requires dimensional analysis based on FEM simulation. This is described in detail in Section 3.4.2.2.

3.3.6 Thermal Stress Method

In the thermal stress method [Nix, 1989; Vinci et al., 1995; Keller et al., 1998], a thin film deposited on a substrate is heated or cooled for loading. The force is applied by the thermal stress originating from the difference in the coefficients of thermal expansion (CTE) between the film and the substrate, as shown in Fig. 3.8. We can impose biaxial compressive and tensile strains in the film by the heat, and the curvature of the thin film/substrate specimen gives us the corresponding film stress. Although stress analysis is required for a precise evaluation, the following Stoney's formula [Stoney, 1909] is often used for convenience:

$$\sigma_{\rm f} = \frac{E_{\rm s}h_{\rm s}^2}{6h_{\rm f}(1-\nu_{\rm s})R} \tag{3.7}$$

Here, $\sigma_{\rm f}$ is the film stress, R is the radius of curvature, $h_{\rm f}$ is the film thickness, $h_{\rm s}$ is the thickness of the substrate, and $E_{\rm s}$ and $\nu_{\rm s}$ are the Young's modulus and Poisson's ratio of the substrate.

Although this is easy, the property at a particular temperature cannot be obtained. In addition, as the heat introduces stress (or strain), experimental and analytical errors might be larger than those introduced during mechanical loading methods.

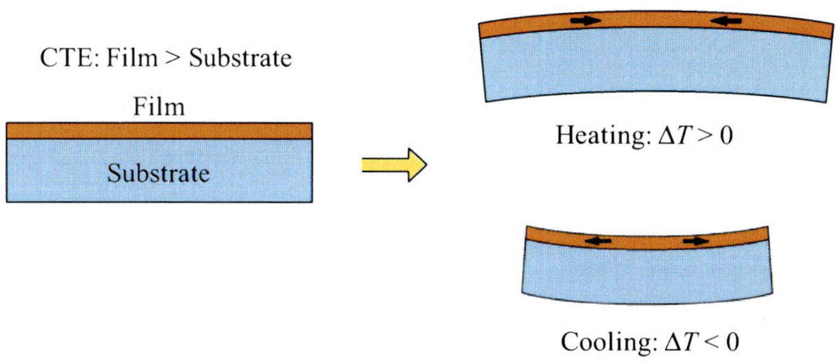

Figure 3.8 Schematic illustrations of a thermal stress test.

3.3.7 Surface Acoustic Wave Method

The velocity of an acoustic wave depends on the Young's modulus, Poisson's ratio, and the density of the materials. Thus, we can evaluate the mechanical properties of a material by analyzing the correlation between them [e.g., Ollendorf et al., 1996]. An acoustic wave is activated on the surface of a thin film by means of a laser, and the velocity is measured by a piezoelectric transducer (Fig. 3.9).

The advantage of this method is that no specific processing is required for specimen preparation. This fundamentally cannot be extended to the evaluation of plastic behavior.

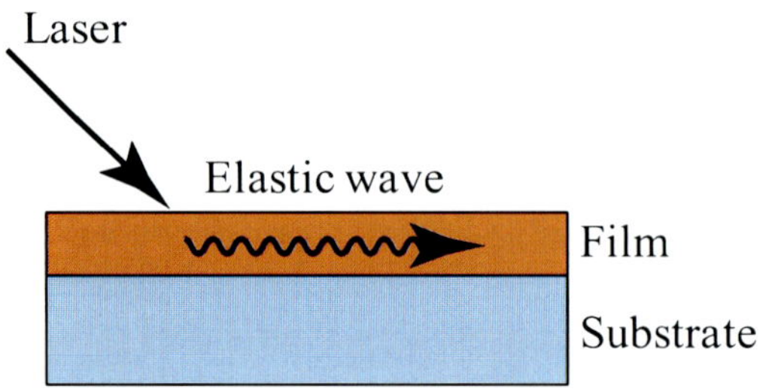

Figure 3.9 Schematic illustrations of a surface wave test.

3.3.8 Vibration Lead Method

The resonant frequency has a specific relation with Young's modulus [e.g., Kiesewetter et al., 1992]. In a cantilever, the relation is in the form

$$E = \frac{48\pi^2 \rho f_n^2 l^4}{\alpha_n^4 d^2} \tag{3.8}$$

where f_n is the resonant frequency, l and d are the length and the thickness of cantilever beam, and α_n is the constant depending on the order of vibration: for the first order $\alpha_1 = 1.8751$, ρ is the density of the thin film. A laser or loudspeaker is used to excite the cantilever structure (Fig. 3.10), and f_n is measured by a noncontact device such as a laser displacement sensor. Because the preparation of a free-standing specimen is difficult, especially for nano-thick films, the method can be extended for a film fixed on a substrate, although the effect of the substrate needs to be considered. This method can evaluate only the elastic property.

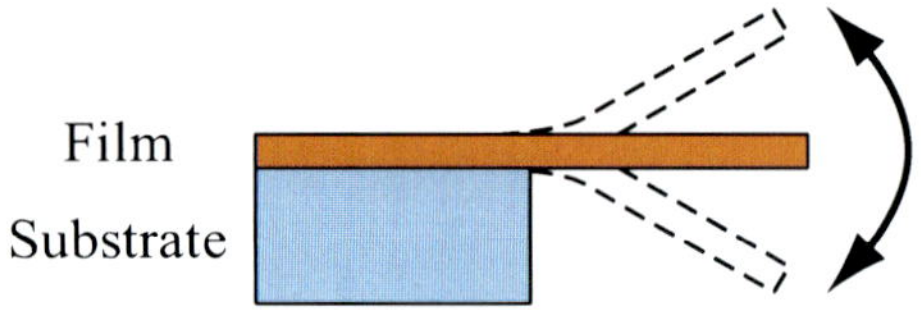

Figure 3.10 Schematic illustrations of the vibration lead method.

3.3.9 Bending Test of Cantilever

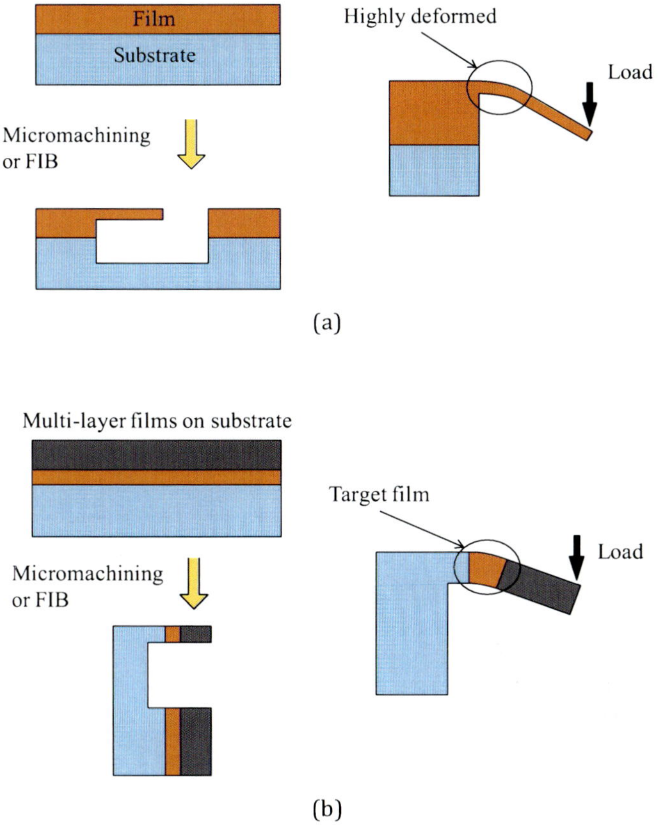

Figure 3.11 Schematic illustration of a cantilever bending test. The cantilever is fabricated from the thin film by micromachining or FIB technique. (a) Cantilever bending. (b) Multi-layer cantilever bending.

The recent developments of FIB and lithography techniques enable us to fabricate a small specimen from a thin film. By loading the small component, we can evaluate the mechanical properties of the film. Figure 3.11 shows the cantilever bend specimens fabricated from thin film on a substrate [Hirakata et al., 2007; Takahashi et al., 2008]. The cantilever consists of a mono-layer (Fig. 3.11a) or multi-layer (Fig. 3.11b). The bending moment is applied by a conventional loading apparatus such as a nanoindenter. Because the specimen is fixed on the substrate, the gripping problem does not have to be considered. The simple bending deformation of the cantilever facilitates the stress analysis. In addition, the advantage of the multilayer method is that it can be applied to ultra-thin films, as described in detail in Section 3.4.4.1. The disadvantage of the method is that specimen preparation by micromachining or FIB is complicated.

3.3.10 Compression Test of Bar

A small bar is fabricated from a thin film on a substrate by micromachining techniques such as lithography and FIB. A compressive load is applied by a flat-ended indenter, as shown in Fig. 3.12. This method can evaluate the deformation behavior in the small region under a uniform compressive stress field, which is different from the indentation test where a large stress gradient occurs around the tip contact region. The size effects on the crystallographic slip behavior are often studied using this method [e.g., Uchic et al., 2004; Greer and Nix, 2006]. The drawback of the method is that specimen preparation by micromachining or FIB is complex.

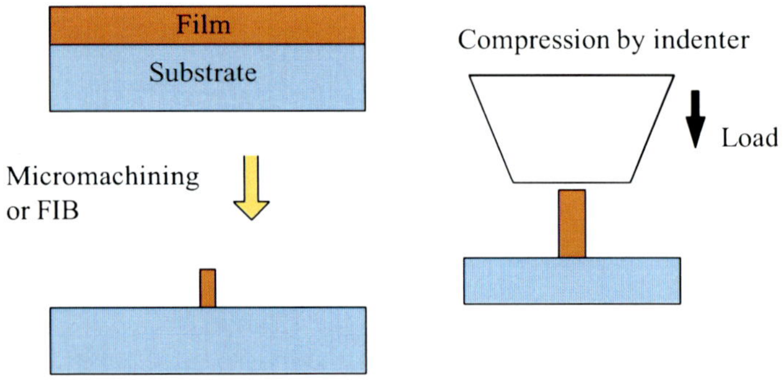

Figure 3.12 Schematic illustration of the compression of a bar. The bar is fabricated from the thin film by micromachining or FIB.

3.4 DEFORMATION PROPERTY

3.4.1 Elasticity

Table 3.2 lists examples of the Young's moduli of various thin films evaluated by the methods explained in Section 3.3. This table only lists the results for thin films of a thickness less than 1000 nm (submicron- or nano-films). The Young's moduli of metallic thin films such as Au, Cu, Al, and Ti are almost the same as or 10 ~ 20% smaller than those of their bulk counterparts [Espinosa et al., 2003a, 2004; Haque and Saif, 2004; Xiang et al., 2002; Yu and Spaepen, 2004; Wei et al., 2007; Haque and Saif, 2002b; Griffin et al., 1992; Tsuchiya et al., 2005; Shan and Sitaraman, 2003]. The 1000-nm-thick Al film exceptionally shows a lower (less than 50%) modulus [Read et al., 2001]. The mechanisms of reduction in the modulus include the existence of defects such as voids and micro-cracks, and the incomplete cohesion of grain boundaries introduced during the vapor deposition process [Xiang, 2005].

Single crystal Si films that are not formed by evaporation show moduli similar to their bulk counterparts [Tsuchiya et al., 2002; Namazu et al., 2002]. On the other hand, other ceramics films deposited by CVD such as poly-Si [Kamiya et al., 2007], SiN [Vlassak and Nix, 1992; Edwards et al., 2004; Zhang et al., 2000], and nanocrystalline diamond [Espinosa et al., 2003b], show a modulus deficit of 10 ~ 30%. This might be due to the effect of film porosity induced during the deposition process.

From these studies, we may conclude that the modulus deficit is more related to the film deposition technique and the ensuing microstructure than to an intrinsic thin-film effect. Therefore, we can increase the Young's modulus up to the same level as the bulk material by recrystallizing the film and thereby reducing the defects by thermal post-processing, such as annealing.

Table 3.2 Examples of experimentally measured Young's modulus for thin films

Material	Thickness, nm	Young's modlus *E*, GPa	*E* of bulk, GPa	Evaluation method	Deposition method	Reference
Au	300, 500, 1000	53–55	82	Free-standing, tensile	EB evaporation	Espinosa et al., 2003a, 2004
Au	200, 350	73, 52	82	Free-standing, tensile	DC sputtering	Haque and Saif, 2004
Cu	200, 1000	125–129	124	Free-standing, tensile	EB evaporation	Espinosa et al., 2004
Cu	885	136	124	Plane strain bulge	Electroplating	Xiang et al., 2002
Cu	100-3000	97.2–108.8	124	Bilayer tensile, on Kapton	EB evaporation	Yu and Spaepen, 2004
Cu	173-998	110–130	124	Plane strain bulge	Thermal evaporation	Wei et al., 2007
Al	30	60.2	69	Free-standing, tensile	Sputtering	Haque and Saif, 2002b
Al	50, 100, 150, 480	63, 68, 70, 70	69	Free-standing, tensile	DC sputtering	Haque and Saif, 2004
Al	200, 1000	65–70	69	Free-standing, tensile	EB evaporation	Espinosa et al., 2004
Al	1000	24.2–30.0	69	Free-standing, tensile	EB evaporation	Read et al., 2001

Material	Thickness, nm	Young's modlus *E*, GPa	*E* of bulk, GPa	Evaluation method	Deposition method	Reference
Al	1000	69.3	69	Bilayer bending, on Si	-	Griffin et al., 1992
Ti	500, 1000	90–100	115–120	Free-standing, tensile	DC sputtering	Tsuchiya et al., 2005
Ti	600	128	115–120	Indentation	DC sputtering	Shan and Sitaraman, 2003
Si	140	158	170, <110>	Free-standing, tensile	Single crystal	Tsuchiya et al., 2002
Si	255	171	170, <110>	Fixed beam bending	Single crystal	Namazu et al., 2002
Poly-Si	760, 920	120–130, 100	170	Free-standing, tensile	LPCVD	Kamiya et al., 2007
SiN	290	222 ± 3 (Bulge), 216 ± 10 (Indentation)	304	Plane strain bulge and indentation	LPCVD	Vlassak and Nix, 1992
SiN	500	257	304	Free-standing tensile and bulge	LPCVD	Edwards et al., 2004
SiN	766	202.57 ± 15.8	304	Fixed beam bending	LPCVD	Zhang et al., 2000
Diamond	600-800	940–970	1100	Free-standing, tensile	PECVD	Espinosa et al., 2003b

3.4.2 Plasticity

Table 3.3 lists examples of the yield stresses of metallic thin films evaluated by the methods shown in Section 3.3. The tensile test and the bulge test are popular for free-standing films, and the indentation test is for films on substrate. As clearly recognized in the results for Au, Cu, and Al films, the material size effect exists in the yield stress; i.e., thinner is stronger [Espinosa et al., 2004; Chasiotis et al., 2007; Emery and Povirk, 2003a, 2003b; Xiang, 2005; Wei et al., 2007; Xiang et al., 2002; Hirakata et al., 2010; Zhao et al., 2007; Haque and Saif 2002a, 2002c, 2001]. Potential hardening mechanisms in these thin films include the restriction of dislocation motion due to the grain boundary (the Hall-Petch relationship), Taylor hardening due to the high dislocation density, the impurities and voids introduced during the deposition process, and the starvation of dislocation sources due to the small volume of thin films. The thickness dependency cannot be explained by classical plasticity theory, and many studies have been devoted to solving the issue using the new models; e.g., the macroscopic model based on continuum mechanics such as the strain gradient plasticity theory [Fleck and Hutchinson, 1993, 1997; Gao et al., 1999; Huang et al., 2000], and the microscopic models such as discrete dislocation dynamics (DDD) [Nicola et al., 2003; Deshpandea et al., 2005]. The detailed explanation of these studies is beyond the scope of this book.

In this section, the elasto-plastic behavior is assumed to be characterized by the simple conventional plasticity theory based on continuum mechanics, and the concepts for determining the material constants of thin films are described below. Plastic properties are indispensable to the understanding of fracture nanomechanics of interfaces, discussed in Chapters 4 and 5.

Table 3.3 Examples of experimentally measured yield stress for thin films

Material	Thickness, nm	Yield stress σ_Y MPa, (hardening exponent n)	Evaluation method	Deposition method	Reference
Au	300	170–220	Free-standing, tensile	EB evaporation	Espinosa et al., 2004
Au	500	140–220	Free-standing, tensile	EB evaporation	Espinosa et al., 2004
Au	1000	55–90	Free-standing, tensile	EB evaporation	Espinosa et al., 2004
Au	500	140–240	Free-standing, tensile	Evaporation	Chasiotis et al., 2007
Au	650	180–295	Free-standing, tensile	Evaporation	Chasiotis et al., 2007
Au	230	236–455	Free-standing, tensile	Evaporation	Emery and Povirk, 2003a, 2003b
Au	250	273–390	Free-standing, tensile	Evaporation	Emery and Povirk, 2003a, 2003b
Au	700	254–474	Free-standing, tensile	Evaporation	Emery and Povirk, 2003a, 2003b
Cu	340	365.6 ± 18.3	Plane strain bulge	Sputtering	Xiang, 2005
Cu	440	346.6 ± 17.3	Plane strain bulge	Sputtering	Xiang, 2005
Cu	610	299.7 ± 15.0	Plane strain bulge	Sputtering	Xiang, 2005
Cu	670	266.1 ± 13.3	Plane strain bulge	Sputtering	Xiang, 2005
Cu	890	250.3 ± 12.5	Plane strain bulge	Sputtering	Xiang, 2005

Material	Thickness, nm	Yield stress σ_Y MPa, (hardening exponent *n*)	Evaluation method	Deposition method	Reference
Cu	173–998	400	Plane strain bulge	Thermal evaporation	Wei et al., 2007
Cu	200	300	Free-standing tensile	EB evaporation	Keller et al., 1996
Cu	885	294.9 ± 1.3	Plane strain bulge	Electroplating	Xiang et al., 2002
Cu	117	773, ($n = 5.9$) ($n' = 0.17$)	Indentation	EB evaporation	Hirakata et al., 2010
Cu	900	164.8, ($n = 4.8$) ($n' = 0.21$)	Indentation	Sputtering	Zhao et al., 2007
Al	100	650	Free-standing, tensile	Sputtering	Haque and Saif 2002c
Al	200	330	Free-standing, tensile	Sputtering	Haque and Saif, 2002a
Al	200	205	Free-standing, tensile	EB evaporation	Espinosa et al., 2004
Al	1000	150	Free-standing, tensile	EB evaporation	Espinosa et al., 2004
Al	100	880	Free-standing, bending	Evaporation	Haque and Saif, 2001
Ti	600	700, ($n = 2.6$) ($n' = 0.35$)	Indentation	DC sputtering	Shan and Sitaraman, 2003
Ni	162	1507, ($n = 10$) ($n' = 0.10$)	Indentation	Thermal evaporation	Hirakata et al., 2010

Our discussion is based on the following power-law constitutive equation.

$$\sigma = \begin{cases} E\varepsilon, & \text{for } \sigma \le \sigma_{\mathrm{Y}} \\ R\varepsilon^{n'} = R\varepsilon^{1/n}, & \text{for } \sigma \ge \sigma_{\mathrm{Y}} \end{cases} \tag{3.9}$$

or

$$\varepsilon = \begin{cases} \sigma/E, & \text{for } \sigma \le \sigma_{\mathrm{Y}} \\ B\sigma^{n} = R^{-n}\sigma^{n}, & \text{for } \sigma \ge \sigma_{\mathrm{Y}} \end{cases} \tag{3.9′}$$

where σ and ε are the von Mises equivalent stress and strain. E, R, and n are the Young's modulus, strength coefficient, and work-hardening exponent, respectively. At the yield stress, we have

$$\varepsilon_{\mathrm{Y}} = \sigma_{\mathrm{Y}}/E = \left(\sigma_{\mathrm{Y}}/R\right)^{n} \tag{3.10}$$

Thus, the constitutive equation includes only three material constants. When E is known, the number of unknown quantities is actually reduced to two, σ_{Y} and n, in the elasto-plastic evaluation.

3.4.2.1 Tensile test of free-standing film

Figure 3.13 shows the stress-strain curve obtained by a tensile test for a Cu thin film with a thickness of 520 nm made by the EB evaporation. The elasto-plastic property is extracted from the stress-strain curve. The Young's modulus of the film is about 106 GPa, and the yield stress, σ_{Y}, which is determined as the one where the curve begins to depart from the linear relation, is evaluated as about 280 MPa. The latter is 4.7 times larger than that of annealed Cu bulk (about 60 MPa [Kuhn and Medlin, 2000]). By fitting the stress-strain curve to Eq. (3.9), we can estimate the hardening exponent n as 1.9(n'=0.53).

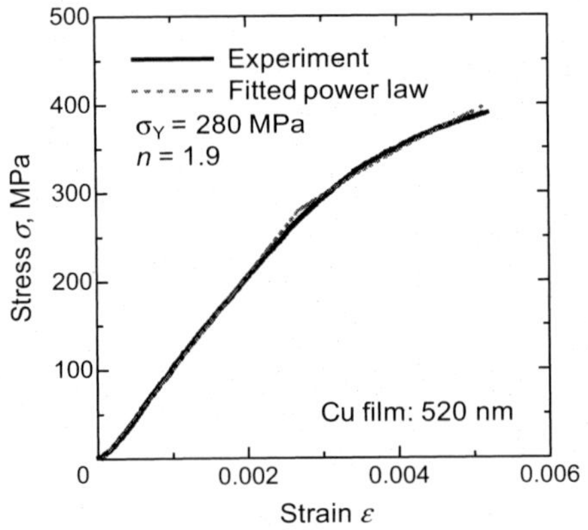

Figure 3.13 Stress-strain relationship in a tensile test for a free-standing Cu film.

3.4.2.2 Indentation test

Unlike the tensile test for free-standing films where the stress-strain relation can be directly evaluated, the indentation test needs numerical analysis to obtain the stress-strain relation from the mechanical response of the specimen. The estimation concept of plastic property (σ_Y and n) is developed using dimensional analysis and FEM simulation [Dao et al., 2001; Chollacoop et al., 2003; Cheng and Cheng, 1998].

In an indentation test for an elasto-plastic material, the relationship between the load P and penetration depth h is generally in the form

$$P = Ch^2 \tag{3.11}$$

where C is the indentation coefficient. The elasto-plastic behavior is characterized by E, ν, σ_Y, n, and the indenter angle ?. The representative stress σ_r defined in Fig. 3.14 can be used instead of σ_Y. In this case, the load P is represented by

$$P = P(h, E, \mathrm{n}, \sigma_r, n, ?) \tag{3.12}$$

Considering the elasticity of the indenter and assuming that the Poisson's ratio is known, the equation becomes

$$P = P(h, E^*, \sigma_r, n, \phi) \tag{3.13}$$

where E^* is the reduced modulus defined in Eq. (3.3).

Applying the Π theorem in the dimensional analysis,

$$P = \sigma_r h^2 \Pi\left(E^*/\sigma_r, n, \phi\right) \tag{3.14}$$

$$C = \frac{P}{h^2} = \sigma_r \Pi\left(E^*/\sigma_r, n, \phi\right) \tag{3.15}$$

The function Π consists of three nondimensional variables. The concrete function of Π is determined by finite element method (FEM) analysis.

For example, the Π function for the Berkovich tip (equivalent cone angle: $\phi = 70.3°$) is shown in Fig. 3.15. In the figure, the vertical axis gives the indentation coefficient C divided by the representative stress σ_r, while the horizontal axis indicates the reduced Young's modulus E^* divided by the representative stress σ_r. Dimensionless function Π is constructed using three different values of ε_r (e.g., $\varepsilon_r = 0.01$, 0.033, and 0.29) and the corresponding σ_r, respectively. A representative plastic strain $\varepsilon_r = 0.033$ can be identified as a strain level where Π is independent of the strain-hardening exponent n. The magnitude of C is obtained from the indentation experiment for a target thin film. Using the Π function at $\varepsilon_r = 0.033$ shown in Fig. 3.15b, the representative stress $\sigma_{0.033}$ is determined by Eq. (3.15). This operation settles the one point $(\sigma_r, \varepsilon_r) = (\sigma_{0.033}, 0.033)$ in the stress-strain relationship shown in Fig. 3.14. In similar way, another point $(\sigma_r, \varepsilon_r)$ can be obtained by the indentation experiment by another indenter with a different face angle ϕ and the corresponding FEM analysis. From the two sets of $(\sigma_r, \varepsilon_r)$, the material constants (σ_Y, n) are uniquely determined.

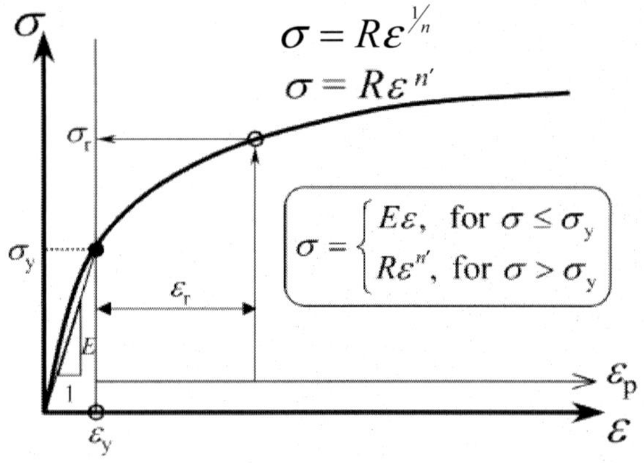

Figure 3.14 Power-law elasto-plastic behavior. Reprinted from Dao, M., et al., "Computational modeling of the forward and reverse problems in instrumented sharp indentation," *Acta Mater.*, 49, pp. 3899–3918, Copyright (2001), with permission from Elsevier.

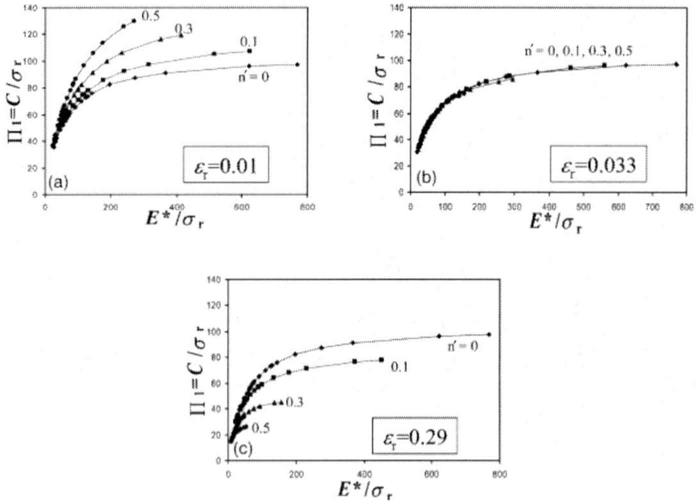

Figure 3.15 The Π function of a Berkovich tip (equivalent cone angle: ? = 70.3) . Reprinted from Dao, M., et al., "Computational modeling of the forward and reverse problems in instrumented sharp indentation," *Acta Mater.*, 49, pp. 3899–3918, Copyright (2001), with permission from Elsevier.

For a bulk and a micron-thick film, the elasto-plastic material constants evaluated by the indentation test have a good agreement with the ones obtained by the conventional tensile test; e.g., 400 μm-thick Ni films [Bucaille et al., 2004]. However, application to thin film is restricted, mainly due to two factors: the difficulty in precise fabrication of the indenter geometry, e.g., tip defects, and the influence of the substrate. It has been shown [Tunvisut et al., 2002] that the latter can be neglected for a ratio of indentation depth to film thickness less than one-tenth.

The above procedure can be extended to a thin film on a substrate, and the elasto-plastic property is evaluated [Zhao et al., 2007]. Figure 3.16 shows the stress-strain curve estimated by the indentation method for a 900 nm-thick Cu film and the result obtained by the bulge method for the free-standing film. Both the results show a good agreement. The elasto-plastic property is evaluated as E = 138.7 GPa, σ_Y = 164.8 MPa and n = 4.8 (n□ = 0.21).

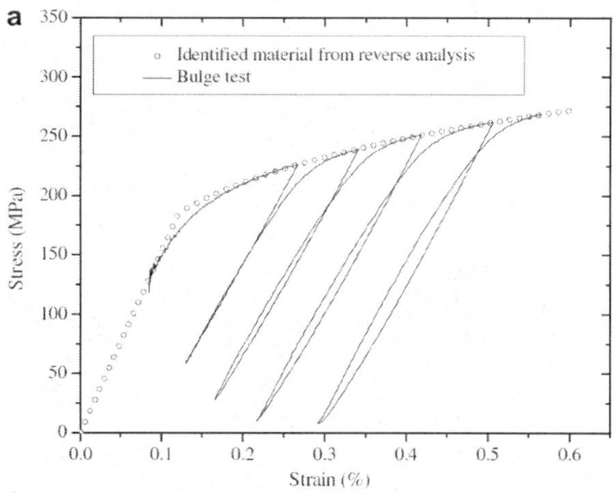

Figure 3.16 Comparison between the plane strain stress–strain curve identified from the indentation test and that measured from the parallel bulge test for a sputter-deposited Cu film on silicon substrate. Reprinted from Zhao, M., et al., "Measuring elastoplastic properties of thin films on an elastic substrate using sharp indentation," Acta Mater., 55, pp. 6260–6274, Copyright (2007), with permission from Elsevier.

Several inverse methods using the indentation experiment and FEM analysis have been proposed for evaluating the elasto-plastic property of thin films [Shan and Sitaraman, 2003; Hirakata et al., 2010], and the properties are examined for submicron-thick films such as Cu, Ni, and Ti films. The indentation method is also extended for the evaluation of anisotropic plasticity, which is often the case in thin films [Yonezu et al., 2009].

3.4.2.3 Bending of cantilever

Figure 3.17 shows a scanning electron microscope (SEM) micrograph of a specimen made of Ta_2O_5 solid thin film with a thickness of 2.8 μm on a Si(100) substrate carved out by means of an FIB [Ohkura et al., 2010]. The specimen is bended by a triangular pyramid chip using atomic force microscopy (AFM). The resolution capabilities of the load and the displacement are 0.1 μN and 0.2 nm, respectively.

Figure 3.18 shows the load-displacement relationship obtained. There is obviously linearity in a low load, and the Young's modulus of thin film is evaluated as 130 GPa using FEM analysis of the gradient of the load-displacement curve. On the other hand, the yield stress σ_Y and the hardening coefficient n are inversely calculated by trial and error using elasto-plastic FEM analysis. The curve fitting by the inverse analysis gives $\sigma_Y = 1970$ MPa and $n = 2.1$ ($n' = 0.48$), as shown in Fig. 3.19, and it has a good coincidence with (σ_r, ε_r) obtained by the indentation test.

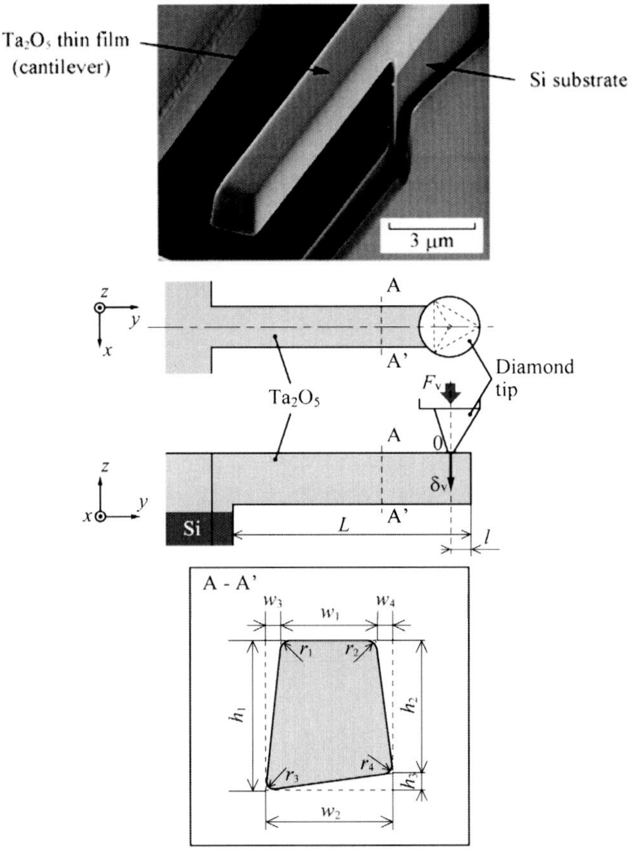

	L	w_1	w_2	w_3	w_4	h_1
Specimen-A	11.23	1.48	1.94	0.24	0.22	2.45
Specimen-B	7.73	1.05	1.23	0.09	0.09	1.77

	h_2	h_3	r_1	r_2	r_3	r_4
Specimen-A	2.13	0.32	0.15	0.15	0.15	0.17
Specimen-B	1.45	0.32	0.18	0.18	0.09	0.15

Figure 3.17 SEM micrograph of a Ta_2O_5 specimen and the loading method.

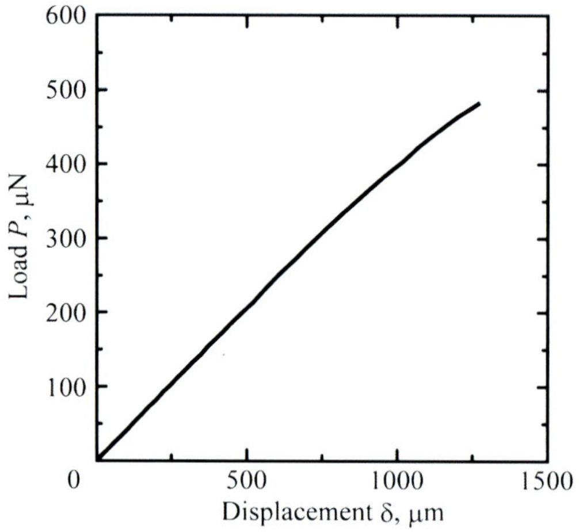

Figure 3.18 Load-displacement curve in a cantilever bend specimen.

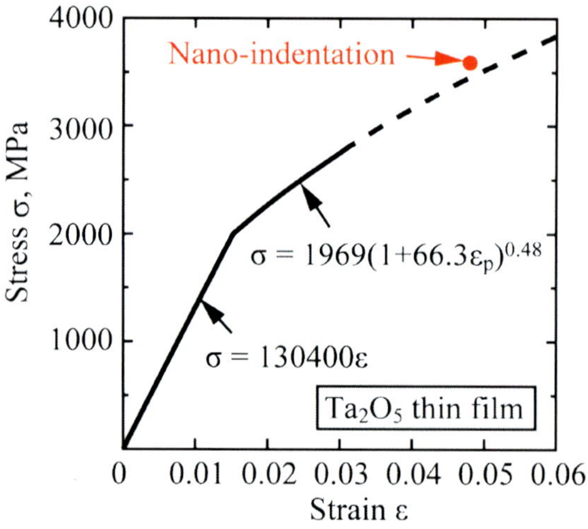

Figure 3.19 Estimated elasto-plastic behavior of a Ta_2O_5 thin film.

3.4.3 Creep

Creep properties are known to be dependent on microstructures [Zhou et al., 1999] and hence the properties of nano films are expected to be vastly different from those of their bulk counterparts. In this section, the power law (Norton's law),

$$\dot{\varepsilon}_c = B_c \sigma^{n_c} \tag{3.16}$$

is assumed for the constitutive equation in creep where B_c and n_c are the material constants. Some of loading methods shown in Section 3.3 are available to determine the material constants.

3.4.3.1 Tensile creep test for a free-standing thin film

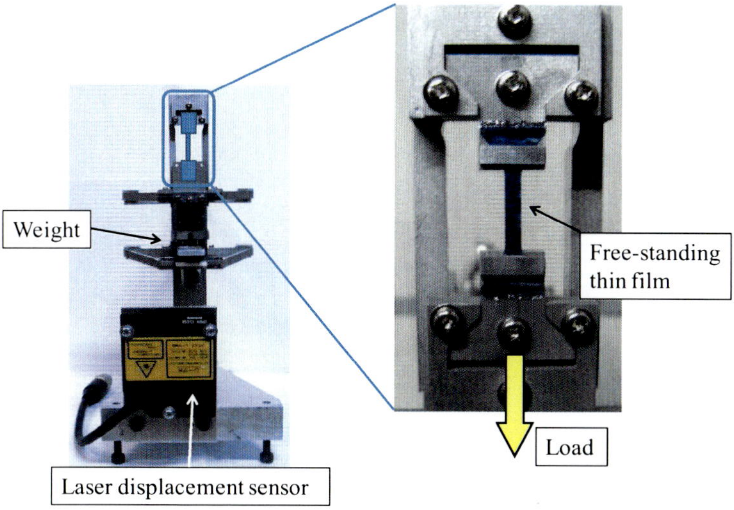

Figure 3.20 Experimental setup for the creep test of a free-standing thin film.

Figure 3.20 shows a tensile creep test for a free-standing Al thin film with a thickness of about 370 nm (grain size 110–120 nm) deposited by EB evaporation. A dead load is applied through a frictionless link mechanism using this experimental setup, and the deformation of the film is measured by a laser displacement sensor.

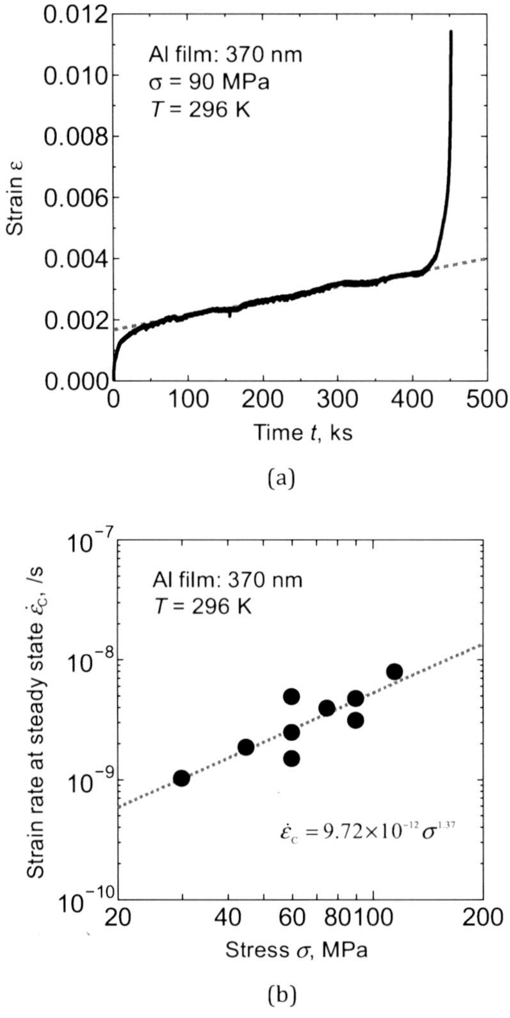

Figure 3.21 Evaluation of the creep property of a 370 nm-thick Al thin film using the tensile creep test. (a) Creep behavior of Al film. (b) Steady state strain rate as a function of applied stress.

Figure 3.21a shows the creep behavior under a constant stress of 90 MPa. The relationship shows a typical creep behavior consisting of transient, steady-state, and accelerated creep regions. Figure 3.21b shows the relationship between the strain rate at the steady state $\dot{\varepsilon}_S$ and the stress σ. The relationship almost meets the Norton law, Eq. (3.16). The creep properties B_c and n_c are estimated as $B_c = 9.72 \times 10^{-12}$ MPa$^{-n_c}$s^{-1} and $n_c = 1.37$, respectively.

3.4.3.2 Multi-layer cantilever bending

The material constants in the creep constitutive equation of 430 nm-thick Sn film evaporated thermally are inversely evaluated from the multi-layer bend experiments and FEM analysis [Hirakata et al., 2008]. The specimen shape is shown in Fig. 3.22. Creep experiments are conducted for three specimens (I–III) under different applied loads F_a. Figure 3.23 shows the relationship between the applied load F_a and the displacement rate under steady state creep $\dot{\delta}_{exp}$. There is a good correlation between F_a and $\dot{\delta}_{exp}$.

The material constants in the steady state creep (B_c, n_c) of Sn are determined by an inverse analysis explained in detail below, assuming that only Sn deforms due to creep. For each specimen, there exist many combinations of (B_c, n_c) that satisfy the experimental result (the displacement rate, $\dot{\delta}_{exp}$) under a magnitude of applied load. However, only one combination can reproduce all of the experimental results in three specimens under the different F_a. This combination is the creep parameter set of Sn. Here, we should be careful that the stress distributes in a cross section of the cantilever.

Figure 3.24 shows the flow chart for determining the (B_c, n_c) relationship for each specimen:

(1) The initial magnitude of n_c is set at unity.

(2) For the n_c value, the displacement rate, $\dot{\delta}_{fem}$, is calculated by the FEM using an arbitrary magnitude of B_c. If the relative error of $\dot{\delta}_{fem}$ to the experimentally measured $\dot{\delta}_{exp}$ is greater than 0.1, another parameter B_c is chosen by trial and error.

(3) If n_c is smaller than 10, process (2) is repeated for the updated $n_c = n_c + 1$.

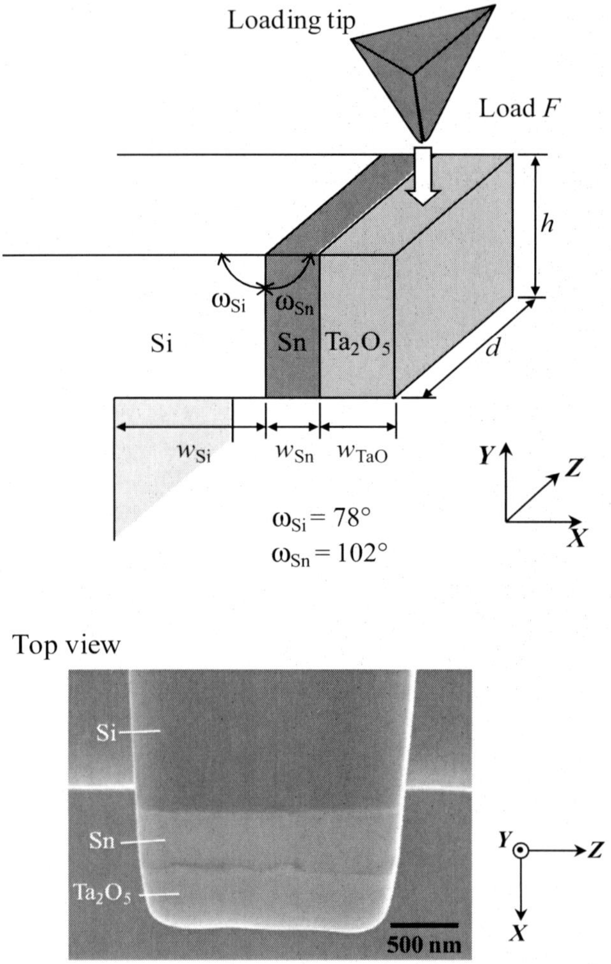

Figure 3.22 Multi-layer cantilever bending specimen for evaluating the creep property of a Sn film. Reprinted from Hirakata, H., et al., "Creep crack initiation at a free edge of an interface between submicron thick elements," *Eng. Fract. Mech.*, 75, pp. 2907–2920, Copyright (2008), with permission from Elsevier.

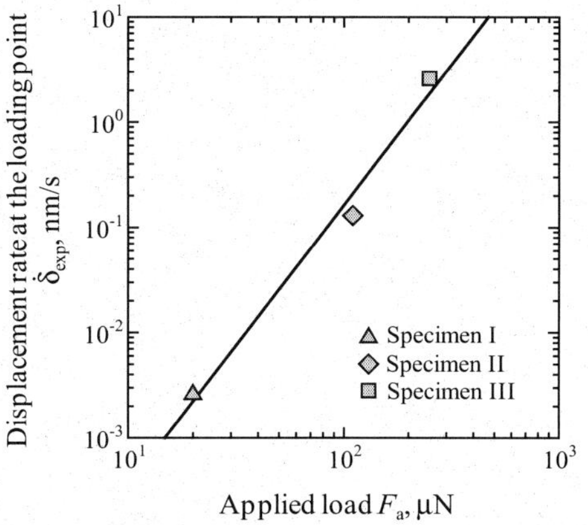

Figure 3.23 Displacement rate as a function of applied force. Reprinted from Hirakata, H., et al., "Creep crack initiation at a free edge of an interface between submicron thick elements," *Eng. Fract. Mech.*, 75, pp. 2907–2920, Copyright (2008), with permission from Elsevier.

Figure 3.24c shows the obtained $B_c - n_c$ relationships for the three specimens (experiments). All of the curves intersect at about $n_c = 3.0$ and $B_c = 1.9 \times 10^{-10}$ MPa$^{-n_c}$s^{-1}. These are the creep parameters of the Sn film.

The creep property of a local region is often evaluated by indentation experiments and analysis [Takagi et al., 2003; Sargent and Ashby, 1992; Li and Ngan, 2004; Bouzakis et al., 2000; Shibutani et al., 2007].

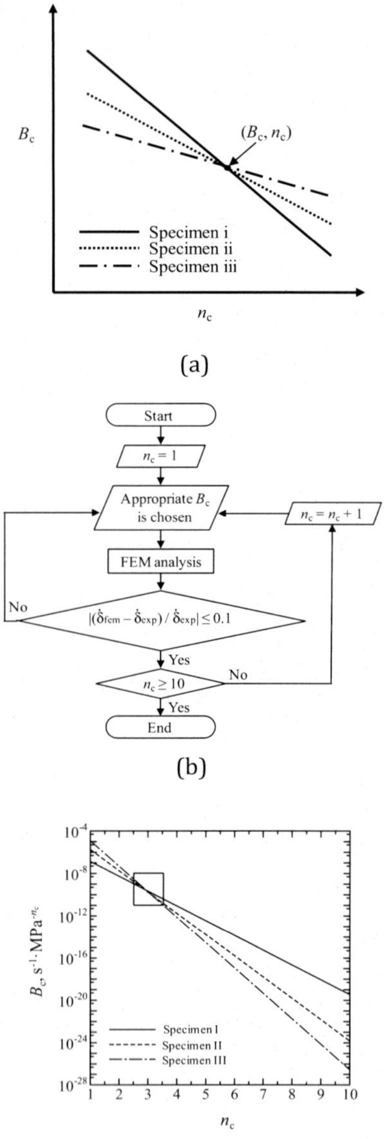

Figure 3.24 Estimation of creep properties B_c and n_c. Reprinted from Hirakata, H., et al., "Creep crack initiation at a free edge of an interface between submicron thick elements," *Eng. Fract. Mech.*, 75, pp. 2907–2920, Copyright (2008), with permission from Elsevier.

3.4.4 Challenge to Evaluation of Elasto-plastic Property in a Nano-element

3.4.4.1 Nano-cantilever bending

When the thickness of a thin film is less than a few tens of nanometers, measurement using the method discussed above becomes difficult. For example, because it is necessary to avoid the effect of the substrate, the indentation test must be conducted to a depth on the order of 10 nm. However, we are already nearly at the accuracy limit of indenter tip fabrication. Thus, the measurable thickness of a thin film is about 100 nm. In the free-standing tensile test method, because it is so difficult to handle the specimen, the reported thinnest film is 30 nm [Haque and Saif, 2002b]. In this section, we introduce a challenging study [Sumigawa, et al., 2010] to determine the elasto-plastic constitutive equation of a 20 nm-thick Cu film.

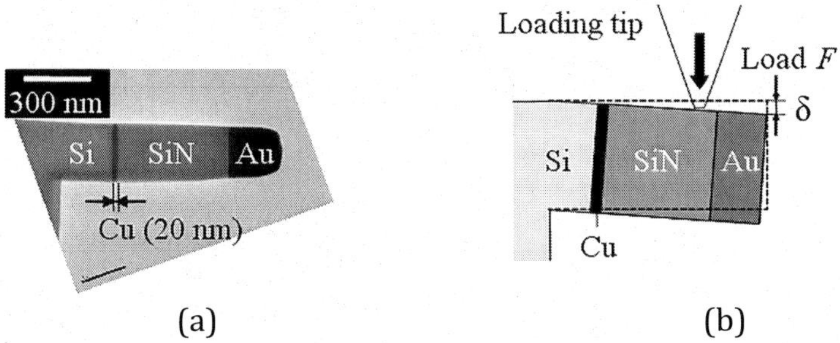

Figure 3.25 TEM micrograph of nano-cantilever bend specimen and the loading style.

The material at issue is a multi-layered thin film of Cu (20 nm thick) and a SiN (500 nm thick) deposited by magnetron sputtering on a Si (100) substrate with a thickness of 500 μm. The cantilever specimen shown in Fig. 3.25a was fabricated by FIB processing. Figure 3.25b shows the schematic illustration of testing where a load is applied to the SiN layer away from the Cu film. The arm amplifies the nonlinear displacement of Cu because Si and SiN are hard and fully elastic. Here, the setup is built into a TEM so that we can make *in situ* observations to measure precisely the displacement and to identify the locations of interface and contact area of the loading chip. The displacement at the arm tip is carefully monitored during the experiment using the TEM images. Varying the ratio of arm length to height, we prepare several specimens to examine the constitutive equation.

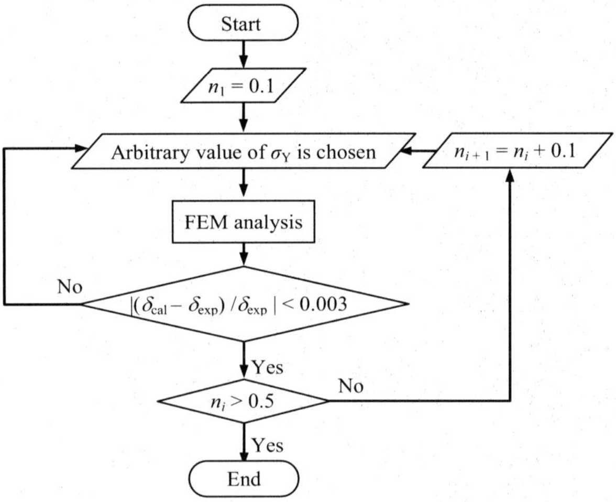

(a)

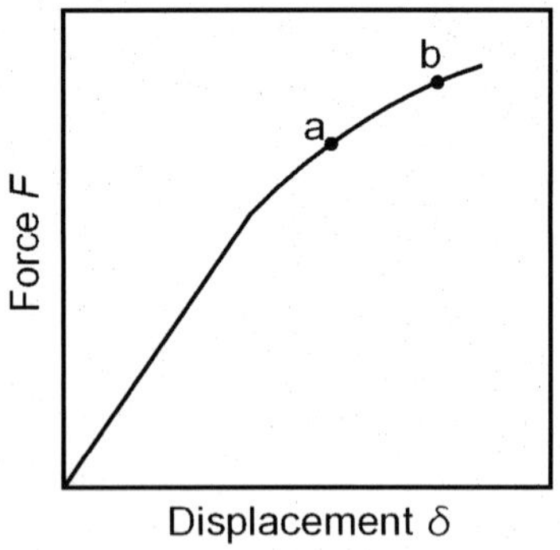

(b)

Figure 3.26 Estimation concept of a plastic property by nano-cantilever bending. (a) Flow chart for determining the σ_Y – n relationship of each specimen. (b) Schematic of force-displacement curve.

The Young's modulus is evaluated from the linear portion of the load-displacement curve by FEM inverse analysis. Then, the inverse-analysis procedure shown in Fig. 3.26a for the nonlinear portion in the load-displacement response can determine the plastic properties (n, σ_Y), which is similar to the procedure in Section 3.4.3.2 for determining creep properties (n_c, B_c). Here, we can use several arbitrary points in the load-displacement curve of one experiment as shown in Fig. 3.26b for a simplification.

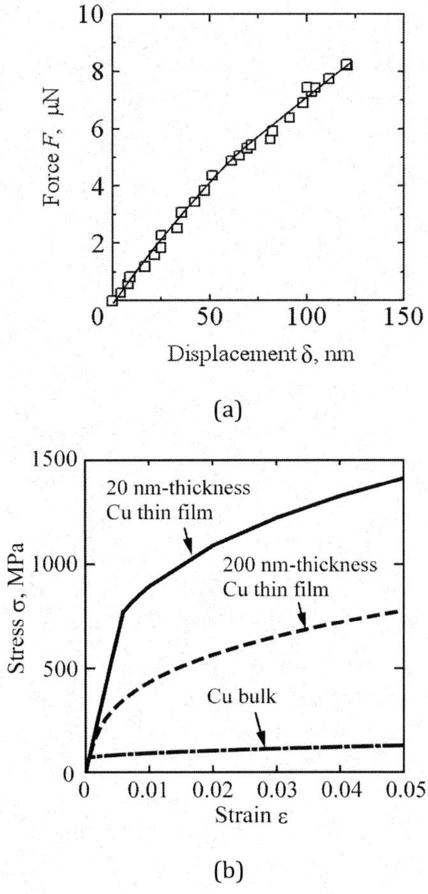

Figure 3.27 Results of the nano-cantilever bend experiment [Sumigawa et al., 2010]. (a) Force-displacement relationship. (b) Stress-strain relationships.

Figure 3.27a shows the obtained relationship between the force, F, and the displacement of the specimen end, δ. The Young's modulus evaluated is E = 129 GPa, which is approximately equal to that of the bulk. The elasto-plastic FEM inverse analysis for several points on the nonlinear portion gives (n, σ_Y) = (2.9, 400 MPa).

Figure 3.27b shows the dependence of σ-ε relations (the constitutive relations) on the film thickness. In the nano-film, the yield stress drastically goes up and the work hardening becomes larger as the thickness decreases.

3.4.4.2 Compression of nano-bar

This section shows another challenge in evaluating deformation of a nano-bar under *in situ* TEM observation. Figure 3.28 shows an Al nano-bar grown by HT-GLAD (high-temperature glancing angle deposition), which has a diameter of about 60 nm. The bar is made of a single crystal, and the longitudinal direction is <110>. The test is conducted in a TEM by the loading system described in Section 5.2.3 (Fig. 5.17).

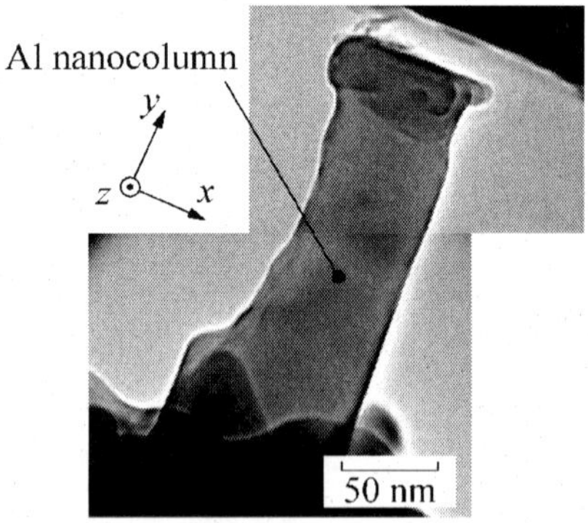

Figure 3.28 Al nano-bar grown by HT-GLAD.

Figure 3.29 shows the load-displacement curve where the nominal stress and strain are indicated on the right and top axes, respectively. The curve is characterized by a sharp yield point (B), an extensive easy glide region (C–G), and a strong work-hardening region (G–I).

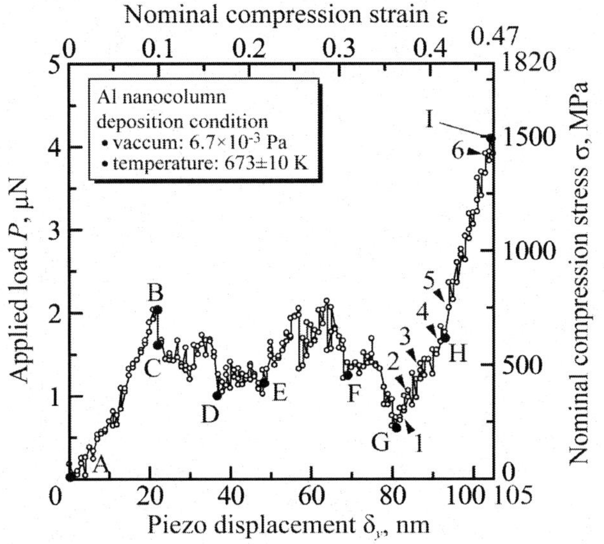

Figure 3.29 The relationship between load and displacement. Nominal compressive stress and strain are indicated by the right and top axes.

Figure 3.30 shows photographs of *in situ* TEM observation, where symbols A–I correspond to the points described in the load-displacement curve (Fig. 3.29). The Al nano-bar deforms without remarkable local distortion until the yielding point (A–B). The top part of the nano-bar gradually becomes thick after yielding. In the easy glide region (C–G), the thickened part spreads over the nano-bar from the top to the bottom. The photographs show the propagation of deformation bands (which might be like the Lüders bands) as well. The specific movement of dislocations, however, is not observed during the deformation until point G. In the work-hardening region (G-I), a huge number of dislocations start to move rapidly at point G and the dislocations continue to multiply after that. Buckling does not occur during the deformation.

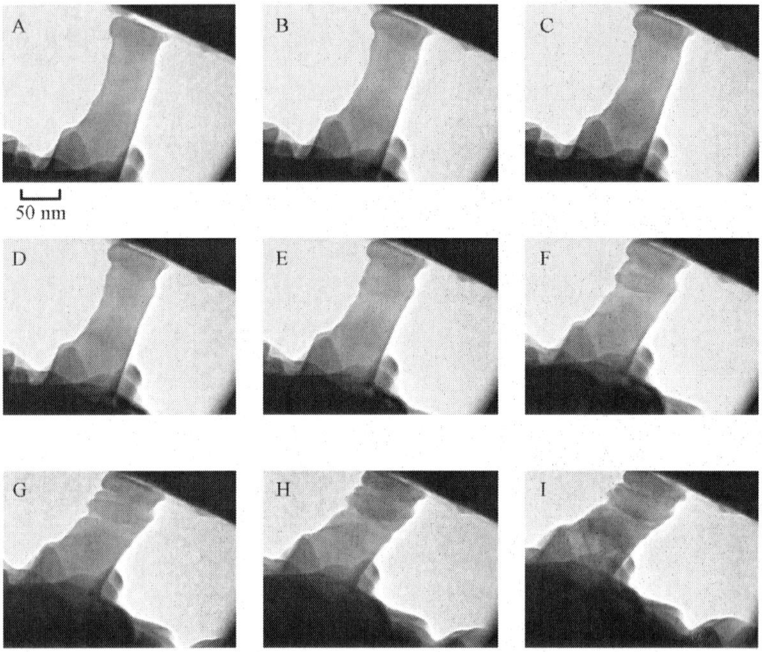

Figure 3.30 Snapshots of the deformation process. The letters A through I correspond to the letters in Fig. 3.29.

The resolved shear stress at the yield point is estimated at about 300 MPa, because the crystallographic orientation parallel to the longitudinal direction of the nano-column is <110>. This is much larger than that of bulk Al single crystal (≈ 1.0 MPa) and is over one-third of the ideal strength (≈ $G/30$ = 850 MPa).

3.5 LIMITATION OF CONTINUUM MECHANICS IN THE DEFORMATION OF THIN FILM

In a single crystal of bulk, it is well known that the initiation of slip deformation is dominated by critical resolved shear stress (Schmid's law). The slip deformation of the polycrystal also follows the critical resolved shear stress of a single crystal. However, the behavior of dislocations in a nano-scale structure is expected to depart from the law. According to dislocation theory, the long-term elastic stress field produced by dislocation is strongly affected by a free surface or a dissimilar interface. For an example, the dislocation near a free surface is attracted to the surface by the mirror force so that the normal stress is zero at the free surface. On the other hand, dislocation near a rigid interface is repulsed by the mirror force so that the displacement is zero at the interface. In a detailed discussion on the elasto-plastic deformation property of nano-scale structures, we must take into account the effects of these boundaries. This suggests that the elasto-plastic deformation cannot be uniquely represented by a conventional equation depending on the boundary conditions (e.g., component shape and boundary) when the size of structure shrinks to nano-scale. The plastic deformation is essentially due to the multiplication, the migration, and the alternative effect of numerous dislocations. The conventional constitutive equation is applicable to the region scale where the dislocations' structure and their dynamic behavior can be statistically averaged in a generalized form. In other words, as the plastic behavior becomes sensitive to the boundary and loading conditions, the constitutive equation, which must be independent of them in terms of conventional continuum mechanics, cannot stand for a very tiny component (e.g., nano-wire with a diameter of a single-digit nanometer). As described in the previous sections, it looks like the conventional constitutive equation is applicable to nano-film (thickness of 10-1000 nm). However, the number of experiments is very limited and further careful investigation is inevitable. For smaller elements, making the equation representing the elasto-plastic behavior is important.

Since it is difficult to extract a universal constitutive equation in a nano-scale material, an analytical method focusing on the dislocation is effective, such as simulations based on discrete dislocation dynamics (DDD) analysis and molecular dynamics. As molecular dynamics are explained in Chapter 7, we briefly describe the methodology of DDD below. It should be noted that these do not give us a unified concept even though they supply *a priori* the deformation behavior at a given boundary condition. In other words, although such simulations can be substituted for an experiment on a nano-element if all the assumptions are correct, they tell us nothing by themselves about the unified law of deformation.

DDD [van der Giessen and Needleman, 1995; Lubarda et al., 1993; Orowan, 1934; Polanyi, 1934; Taylor 1934] is a simulation method in which the movement of each dislocation embedded in a continuum body is examined by considering the force acting on it under a certain boundary condition. Let us consider a finite body with discrete dislocations, where the force boundary and the displacement boundary are defined as S_t and S_u (the original problem) (Fig. 3.31a). The stress distribution of the finite body and the shear stress acting on dislocation are obtained as follows.

(1) The original problem is divided into Problem A (Fig. 3.31b) and Problem B (Fig. 3.31c).

(2) The stress and displacement fields produced by a dislocation in an infinite field are given by an analytical solution. For example, the stress caused by an edge dislocation at the origin parallel to the x axis is [Burgers, 1939]

$$\sigma_{xx} = -\frac{\mu b}{2\pi(1-\nu)} \frac{y(3x^2+y^2)}{(x^2+y^2)^2} \tag{3.17}$$

$$\sigma_{yy} = \frac{\mu b}{2\pi(1-\nu)} \frac{y(x^2-y^2)}{(x^2+y^2)^2} \tag{3.18}$$

$$\sigma_{xy} = \sigma_{yx} = \frac{\mu b}{2\pi(1-\nu)} \frac{x(x^2-y^2)}{(x^2+y^2)^2} \tag{3.19}$$

where μ is a shear modulus. The displacements are

$$\varepsilon_{xx} = -\frac{b}{2\pi} \frac{y\left\{(2\lambda+3\mu)x^2+\mu y^2\right\}}{(\lambda+2\mu)(x^2+y^2)^2} \tag{3.20}$$

$$\varepsilon_{yy} = \frac{b}{2\pi} \frac{y\left\{(2\lambda+\mu)x^2-\mu y^2\right\}}{(\lambda+2\mu)(x^2+y^2)^2} \tag{3.21}$$

$$\varepsilon_{xy} = \varepsilon_{yx} = \frac{b}{4\pi(1-\nu)} \frac{x(x^2-y^2)}{(x^2+y^2)^2} \tag{3.22}$$

The superimposition of contribution by all dislocations produces the fields in Problem A.

(3) The displacement on S_u and the traction vector on S_t are specified in Problem A.

(4) On the basis of the load and constraint conditions of the original analytical model, the solution of Problem B is obtained using FEM analysis.

(5) By superposing Problems A and B, we obtain the stress field of the original problem.

(6) Finally, we obtain the resolved shear stress acting on each dislocation.

The velocity of dislocation motion is decided by the resolved shear stress. For example, the glide velocity v is linearly related to the shear stress τ as

$$v = \frac{\tau}{B'} b \tag{3.23}$$

where b is the Burgers vector and the drug coefficient B? is determined by experiments or MD simulations [Nakatani, 1997]. The disappearance and nucleation of dislocation can be introduced by an appropriate criterion.

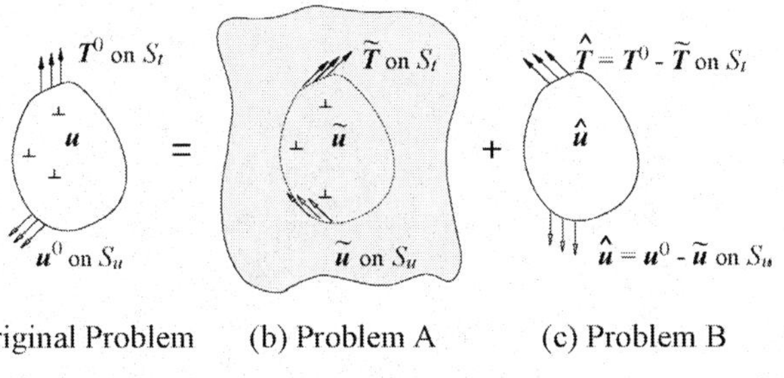

Figure 3.31 Decomposition and superposition of a problem.

By integrating the velocity, we can simulate the time-evolution of dislocation distribution. The disappearance and nucleation of dislocation (e.g., at the free surface) can be introduced by an appropriate criterion.

Research on application of DDD to a thin film under stress is ongoing [Han et al. 2006]. It indicates that the specific deformation layer exists near the free edge and the dissimilar interface, and that the layer greatly affects the deformation behavior of a nano-element.

3.6 FRACTURE OF THIN FILMS

Table 3.4 lists examples of tensile strength and fracture toughness of thin films for several metals and ceramics. Thin films made of pure metal tend to show macroscopically brittle fracture without large elongation, unlike their bulk counterparts. This trend becomes clearer in the thinner films.

In order to clarify the mechanism of the unique fracture behavior of thin film, tensile experiments for free-standing nano-films have been conducted under *in situ* SEM and TEM observation [Haque and Saif, 2004]. Figure 3.32 shows the tensile test instrument for a free-standing 100 nm-thick aluminum film bonded to a force sensor beam at one end and to a set of supporting beams at the other end. Displacements are imposed on one end of the stage and are transmitted to the force sensor beam through the free-standing specimen. The displacement sensors (in the form of two marker gaps, F and D) read the specimen elongation. The force on the specimen is given by the spring constant and the displacement (marker F) of the force sensor beam.

Figure 3.33 shows stress–strain diagrams for three specimens tested in TEM and one specimen in a scanning electron microscope that highlight the consistency of the experiment's results. Also shown are TEM images of a sample at different levels of stress and strain and during failure. The top right TEM image shows the film after failure with a crack normal to the loading direction. The TEM images show neither any dislocation activity within the grains nor any generation, even at high stress. The specimens behave as macroscopically brittle material and fail at the stress of 680 MPa. At 680 MPa, a crack is initiated from the edge and propagates both through the grain boundaries and within the grains. Dislocation activity is limited near the crack tip, and this is why the brittle fracture occurs even in intrinsically ductile materials such as Al.

Table 3.4 Example of experimentally measured tensile strength and fracture toughness of thin films

Material	Thickness, nm	Tensile strength, MPa	Evaluation method	Deposition method	Reference
Au	800	346 ± 7	Free-standing, tensile	Evaporation	Emery and Povirk, 2003a
Cu	200	380	Free-standing, tensile	EB evaporation	Keller et al., 1996
Al	30	325	Free-standing, tensile	Sputtering	Haque and Saif, 2002b
Al	1000	151 ± 19.8	Free-standing, tensile	EB evaporation	Read et al., 2001
Ti	500, 1000	650—800	Free-standing, tensile	DC sputtering	Tsuchiya et al., 2005
Si	140	1400—4900	Free-standing, tensile	Single crystal	Tsuchiya et al., 2002
Poly-Si	760	1800—2300	Free-standing, tensile	LPCVD	Kamiya et al., 2007
Poly-Si	920	1700	Free-standing, tensile	LPCVD	Kamiya et al., 2007

Material	Thickness, nm	Tensile strength, MPa	Evaluation method	Deposition method	Reference
SiN	500	5800	Free-standing, tensile	LPCVD	Edwards et al., 2004
SiN	766	12260 ± 1690 (bending strength)	Fixed-beam bending	LPCVD	Zhang et al., 2000
Material	**Thickness, nm**	**Fracture toughness, $MPam^{-0.5}$**	**Evaluation method**	**Deposition method**	**Reference**
Au	300	0.45	Free-standing, tensile	FIB	Hosokawa et al., 2008
Al	200	0.515	Free-standing, tensile	Sputtering	Hosokawa et al., 2008
Diamond	800	4.6 ± 0.18	Free-standing, tensile	PECVD	Espinosa and Peng, 2005

Using a similar testing setup, tensile tests for notched films have been conducted to evaluate the fracture toughness [Hosokawa et al., 2008]. The notch, which acts as pre-crack, is introduced by FIB. As listed in Table 3.4, thin metallic films such as Au (300 nm thick) and Al (200 nm thick) show much smaller toughness (roughly two orders of magnitude smaller) than those of their bulk counterparts. Because the crack in the thin film is under the strong plain stress condition, mechanical constraint for the plastic deformation is small compared with that under the plain strain condition. However, the dislocation motion is restricted in the thin film, as mentioned before, leading to the high stress level near the crack tip. Hence, the thin films show apparently weak fracture toughness.

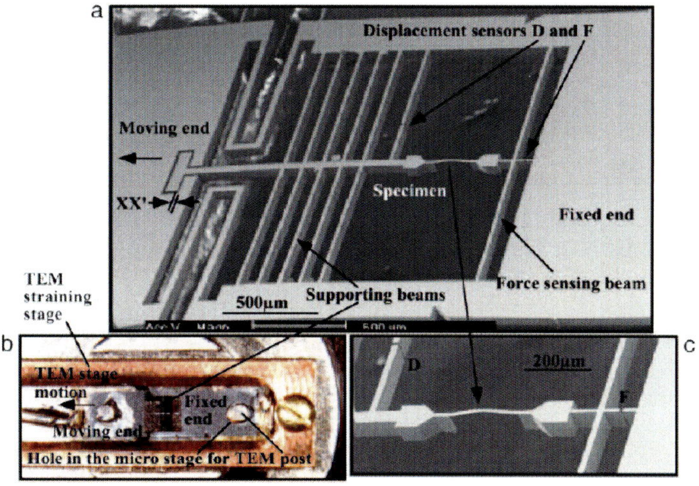

Figure 3.32 Experimental setup of the in situ tensile test for a free-standing aluminum thin-film specimen. Reprinted with permission from Haque, M. A., and Saif, M. T. A., "Deformation mechanisms in free-standing nanoscale thin films: a quantitative in situ transmission electron microscope study," *Proc. Nat. Acad. Sci.*, 101, pp. 6335–6340. Copyright (2004) National Academy of Sciences, U.S.A.

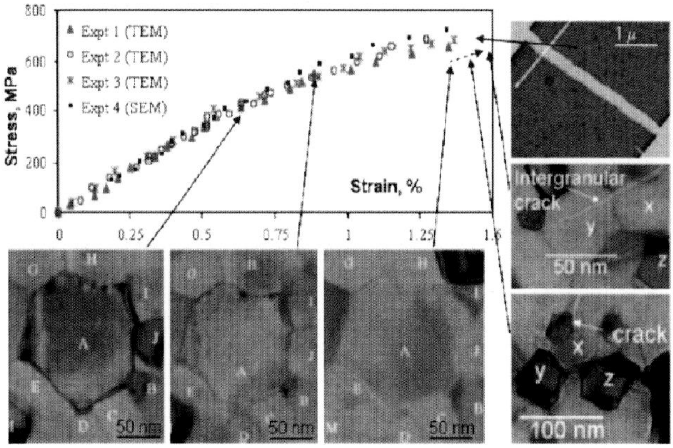

Figure 3.33 The results of the *in situ* TEM tests for 100-nm-thick free-standing aluminum films. Reprinted with permission from Haque, M. A., and Saif, M. T. A., "Deformation mechanisms in free-standing nanoscale thin films: a quantitative in situ transmission electron microscope study," *Proc. Nat. Acad. Sci.*, 101, pp. 6335–6340. Copyright (2004) National Academy of Sciences, U.S.A.

In ceramic films, fracture behavior is relatively similar to that of their bulk counterparts. Since a lot of small defects dominate the fracture strength, statistical concepts such as Weibull analysis are important to evaluating it [Espinosa et al., 2003b].

Studies of the fracture of thin films are very few compared with studies of deformation. Future research in this field should explore various mechanical laws of fracture strength, including subcritical crack growth such as fatigue, creep, and environment-assisted cracking.

References

Bouzakis, K. D., Lontos, A., Vidakis, N., David, K., and Kechagias, V. (2000) Determination of creep behavior of monolayer thick plasma sprayed coatings, by means of the impact test and an analytical FEM supported evaluation procedure, *Thin Solid Films*, 377–378, pp. 373–381.

Bucaille, J. L., Stauss, S., Schwaller, P., and Michler, J. (2004) A new technique to determine the elastoplastic properties of thin metallic films using sharp indenters, *Thin Solid Films,* 447–448, pp. 239–245.

Burgers, J. M. (1939) Some considerations on the fields of stress connected with dislocations in a regular crystal lattice I, *Proceedings of the koninklijkenedlandese akadmie van wetenschappen,* 42, pp. 293–325.

Chasiotis, I., Bateson, C., Timpano, K., McCarty, A. S., Barker, N. S., and Stanec, J. R. (2007) Strain rate effects on the mechanical behavior of nanocrystalline Au films I, *Thin Solid Films,* 515, pp. 3183–3189.

Chen, X., and Vlassak, J. J. (2001) Numerical study on the measurement of thin film mechanical properties by means of nanoindentation, *J. Mater. Res.,* 16, pp. 2974–2982.

Cheng, Y. T., and Cheng, C. M. (1998) Scaling approach to conical indentation in elastic-plastic solids with work hardening, *J. Appl. Phys.,* 84, pp. 1284–1291.

Chollacoop, N., Dao, M., and Suresh, S. (2003) Depth-sensing instrumented indentation with dual sharp indenters, *Acta Mater.,* 51, pp. 3713–3729.

Dao, M., Chollacoop, N., Van Vliet, K. J., Venkatesh, T. A., and Suresh, S. (2001) Computational modeling of the forward and reverse problems in instrumented sharp indentation, *Acta Mater.,* 49, pp. 3899–3918.

Deshpandea, V. S., Needleman, A., and Van der Giessen, E. (2005) Plasticity size effects in tension and compression of single crystals, *J. Mech. Phys. Solids,* 53, pp. 2661–2691.

Eckertova, L. (1986) *Physics of Thin Films, 2nd edn,* Plenum Press, New York.

Edwards, R. L., Coles, G., and Sharpe, Jr., W. N. (2004) Comparison of tensile and bulge tests for thin-film silicon nitride, *Exp. Mech.,* 44, pp. 49–54.

Emery, R. D., and Povirk, G. L. (2003a) Tensile behavior of free-standing gold films. Part I. Coarse-grained films, *Acta Mater.,* 51, pp. 2067–2078.

Emery, R. D., and Povirk, G. L. (2003b) Tensile behavior of free-standing gold films. Part II. Fine-grained films, *Acta Mater.,* 51, pp. 2079–2087.

Espinosa, H. D., Prorok, B. C., and Fischer, M. (2003a) A methodology for determining mechanical properties of freestanding thin films and MEMS materials, *J. Mech. Phys. Solids,* 51, pp. 47–67.

Espinosa, H. D., Peng, B., Prorok, B. C., Moldovan, N., Auciello, O., Carlisle, J. A., Gruen, D. M., and Mancini, D. C. (2003b) Fracture strength of ultrananocrystalline diamond thin films—identification of Weibull parameters, *J. Appl. Phys.,* 94, pp. 6076–6084.

Espinosa, H. D., Prorok, B. C., and Peng, B. (2004) Plasticity size effects in free-standing submicron polycrystalline FCC films subjected to pure tension, *J. Mech. Phys. Solids*, 52, pp. 667–689.

Espinosa, H. D., and Peng, B. (2005) A new methodology to investigate fracture toughness of freestanding MEMS and advanced materials in thin film form, *J. Microelectromech. Syst.*, 14, pp. 153–159.

Florando, J. N., and Nix, W. D. (2005) A microbeam bending method for studying stress-strain relations for metal thin films on silicon substrates, *J. Mech. Phys. Solids*, 53, pp. 619–638.

Fleck, N. A., and Hutchinson, J. W. (1993) A phenomenological theory for strain gradient effects in plasticity, *J. Mech. Phys. Solids*, 41, pp. 1825–1857.

Fleck, N. A., and Hutchinson, J. W. (1997) Strain gradient plasticity, *Adv. Appl. Mech.*, 33, pp. 295–361.

Freund, L. B., and Suresh S. (2003) *Thin Film Materials, Stress, Defect Formation and Surface Evolution*, Cambridge University Press.

Gao, H., Huang, Y., Nix, W. D., and Hutchinson, J. W. (1999) Mechanism-based strain gradient plasticity—I. Theory, *J. Mech. Phys. Solids*, 47, pp. 1239–1263.

Greer, J. R., and Nix, W. D. (2006) Nano-scale gold pillars strengthened through dislocation starvation, *Phys. Rev. B*, 73, p. 245410.

Griffin, A. J., Brotzen, F. R., and Dunn, C. F. (1992) Mechanical testing of thin metallic films, *Thin Solid Films*, 220, pp. 265–270.

Han, C.-S., Hartmaier, A., Gao, H., and Huang, Y. (2006) Discrete dislocation dynamics simulations of surface induced size effects in plasticity. *Mater. Sci. Eng. A*, 415, pp. 225–233.

Haque, M. A., and Saif, M. T. A. (2001) Microscale materials testing using MEMS actuators, *J. Microelectromech. Syst.*, 10, pp. 146–152.

Haque, M. A., and Saif, M. T. A. (2002a) In situ SEM tensile testing of nano-scale specimen in SEM and TEM , *Exp. Mech.*, 42, pp. 123–128.

Haque, M. A., and Saif, M. T. A. (2002b) Mechanical behavior of 30–50 nm thick aluminum films under uniaxial tension, *Scripta Mater.*, 47, pp. 863–867.

Haque, M. A., and Saif, M. T. A. (2002c) Application of MEMS forse sensors for *in situ* mechanical characterization of nano-scale think films in SEM and TEM, *Sens. Actuators*, A 97-98, pp. 239-345.

Haque, M. A., and Saif, M. T. A. (2004) Deformation mechanisms in free-standing nano-scale thin films: a quantitative in situ transmission electron microscope study, *Proc. Nat. Acad. Sci.,* 101, pp. 6335–6340.

Hay, J. C., Bolshakov, A., and Pharr, G. M. (1999) A critical examination of the fundamental relations used in the analysis of nanoindentation data, *J. Mater. Res.,* 14, pp. 2296–2305.

Hirakata, H., Takahashi, Y., Truong, D. V., and Kitamura, T. (2007) Role of plasticity on interface crack initiation at free edge and its propagation in nanostructures, *Int. J. Fract.,* 145, pp. 261–271.

Hirakata, H., Hirako, T., Takahashi, Y., Matsuoka, Y., and Kitamura, T. (2008) Creep crack initiation at a free edge of an interface between submicron thick elements, *Eng. Fract. Mech.,* 75, pp. 2907–2920.

Hirakata, H., Yamada, T., Nobuhara, Y., Yonezu, A., and Minoshima, K. (2010) Hydrogen effect on fracture toughness of thin film/substrate interfaces, *Eng. Fract. Mech.*, 77, pp. 803–818.

Hosokawa, H., Desai, A. V., and Haque, M. A. (2008) Plane stress fracture toughness of freestanding nano-scale thin films, *Thin Solid Films,* 516, pp. 6444–6447.

Huang, Y., Gao, H., Nix, W. D., and Hutchinson, J. W. (2000) Mechanism-based strain gradient plasticity—II. Analysis, *J. Mech. Phys. Solids*, 48, pp. 99–128.

Kamiya, S., Kuypers, J. H., Trautmann, A., and Ruther, P. (2007) Process temperature-dependent mechanical properties of polysilicon measured using a novel tensile test structure, *J. Microelectromech. Syst.,* 16, pp. 202–212.

Keller, R. R., Phelps, J. M., and Read, D. T. (1996) Tensile and fracture behavior of freestanding copper films, *Mater. Sci. Eng. A,* 214, pp. 42–52.

Keller, R. M., Baker, S. P., and Arzt, E. (1998) Quantitative analysis of strengthening mechanisms in thin Cu films: effects of film thickness, grain size, and passivation, *J. Mater. Res.,* 13, pp. 1307–1317.

Kiesewetter, L., Zhang, J. M., Houdeau, D., and Steckenborn, A. (1992) Determination of Young's moduli of micromechanical thin films using the resonance method, *Sens. Actuators, A* 35, pp. 153–159.

Kuhn, H., and Medlin, D. (2000) *ASM Handbook, vol. 8: Mechanical Testing and Evaluation*, ASM International, Ohio, p. 103.

Li, H., and Ngan, A. H. W. (2004) Size effects of nanoindentation creep, *J. Mater. Res.,* 19, pp. 513–522.

Lubarda, V. A., Blume, J. A., and Needleman, A. (1993) An analysis of equilibrium dislocation distributions, *Acta Metall. Mater.*, 41, pp. 625–642.

Nakatani, A., Kitagawa, H., and Nakatani, K. (1997) Molecular dynamics study of relationship between shear stress and velocity for discrete dislocation dynamics, *J. S oc. Mat. Sci. Japan,* 46, pp. 619–624 (in Japanese).

Namazu, T., Isono, Y., and Tanaka, T. (2002) Plastic deformation of nanometric single crystal silicon wire in AFM bending test at intermediate temperatures, *J. Microelectromech. Syst.,* 11, pp. 125–135.

Nicola, L., Van der Giessen, E., and Needleman, A. (2003) Discrete dislocation analysis of size effects in thin films, *J. Appl. Phys.,* 93, pp. 5920–5928.

Nix, W. D. (1989) Mechanical properties of thin films, *Metallurgical Trans. A,* 20, pp. 2217–2245.

Ohkura, Y., Sumigawa, T., Shuhei I, and Kitamura, T. (2010) Evaluation on mechanical property of Ta_2O_5 thin film by microscale bending test, *Trans. JSME*, submitted.

Oliver, W. C., and Pharr, G. M. (1992) An improved technique for determining hardness and elastic modulus using load and displacement sensing indentation experiments, *J. Mater. Res.,* 7, pp. 1564–1583.

Ollendorf, H., Schnaider, D., Schwarz, T., Kirchhoff, G., and Mucha, A. (1996) A comparative study of the mechanical properties of TiN coatings using the non-destructive surface acoustic wave method, scratch test and four-point bending test, *Surf. Coat. Technol.,* 84, pp. 458–464.

Orowan, E. (1934) Zur Kristallplastizität. III Über den Mechanismus des Gleitvorganges, *Zeitschrift für Physik A Hadrons and Nuclei*, 89, pp. 634–659 (in German).

Polanyi, M. (1934) Über eine Art Gitterstörung, die einen Kristall plastisch machen könntes, *Zeitschrift für Physik A Hadrons and Nuclei*, 89, pp. 660–664 (in German).

Read, D. T., Cheng, Y. W., Keller, R. R., and McColskey, J. D. (2001) Tensile properties of free-standing aluminum thin films, *Scripta Mater.,* 45, pp. 583- 589.

Saha, R., and Nix, W. D. (2002) Effects of the substrate on the determination of thin film mechanical properties by nanoindentation, *Acta Mater.,* 50, pp. 23- 38.

Sargent, P. M., and Ashby, M. F. (1992) Indentation creep, *Mater. Sci. Technol.,* 8, pp. 594–601.

Shan, Z., and Sitaraman, S. K. (2003) Elastic–plastic characterization of thin films using nanoindentation technique, *Thin Solid Films,* 437, pp. 176–181.

Shibutani, T., Yu, Q., and Shiratori, M. (2007) A study of deformation mechanism during nanoindentation creep in tin-based solder balls, *J. Electron. Packag.,* 129, pp. 71–75.

Stoney, G. G. (1909) The tension of metallic films deposited by electrolysis, *Proc. R. Soc. London A*, 82, pp. 172–175.

Sumigawa, T., Shishido, T., Murakami, T., Iwasaki, T., and Kitamura, T. (2010) Evaluation of plastic deformation property of copper nano-film by nano-scale cantilever specimen, *Thin Solid Films,* submitted.

Tabata, O., Kuwahata, K., Sugiyama, S., and Igarashi, I. (1989) Mechanical property measurement of thin films using load deflection of composite membranes, *Sens. Actuators,* 20, pp. 135–141.

Takagi, H., Dao, M., Fujiwara, M., and Otsuka, M. (2003) Experimental and computational creep characterization of Al–Mg solid-solution alloy through instrumented indentation, *Philos. Mag,.* 11, pp. 3959–3976.

Takahashi, Y., Hirakata, H., and Kitamura, T. (2008) Quantitative evaluation of plasticity of a ductile nano-component, *Thin Solid Films,* 516, pp. 1925–1930.

Taylor, G. I. (1934) The mechanism of plastic deformation of crystals. Part I. Theoretical, *Proc. R. Soc. London A*, 145, pp. 362–387.

Tsuchiya, T., Shikida, M., and Sato, K. (2002) Tensile testing system for sub-micrometer thick films, *Sens. Actuators A,* 97–98, pp. 492–496.

Tsuchiya, T., Hirata, M., and Chiba, N. (2005) Influence of nanometre-sized notch and water on the fracture behaviour of single crystal silicon microelements, *Thin Solid Films,* 484, pp. 245–250.

Tsui, T. Y., Vlassak, J. J., and Nix, W. D. (1999) Indentation plastic displacement field: Part I. The case of soft films on hard substrates, *J. Mater. Res.,* 14, pp. 2196–2203.

Tunvisut, K., Busso, E. P., O'dowd, N. P., and Brantner, H. P. (2002) Determination of the mechanical properties of metallic thin films and substrates from indentation tests, *Philos. Mag,.* 82, pp. 2013–2029.

Uchic, M. D., Dimiduk, D. M., Florando, J. N., and Nix, W. D. (2004) Sample dimension influence strength and crystal plasticity, *Science,* 305, pp. 986–989.

van der Giessen, E., and Needleman, A. (1995) Discrete dislocation plasticity—a simple planar model, *Model. Simul. Mater. Sci. Eng.*, 3, pp. 689–735.

Vinci, R. P., Zielinski, E. M., and Bravman, J. C. (1995) Thermal strain and stress in copper thin films, *Thin Solid Films,* 262, pp. 142–153.

Vlassak, J. J., and Nix, W. D. (1992) A new bulge test technique for the determination of Young's modulus and Poisson's ratio of thin films, *J. Mater. Res.,* 7, pp. 3242–3249.

Wei, X., Lee, D., Shim, S., Chena, X., and Kysara, J. W. (2007) Plane-strain bulge test for nanocrystalline copper thin film, *Scripta Mater.,* 57, pp. 541–544.

Xiang, Y., Chen, X., and Vlassak, J. J. (2002) The mechanical properties of electroplated Cu thin films measured by means of the bulge test technique, *Mat. Res. Soc. Symp. Proc.,* 695, pp. L4.9.1–L4.9.6.

Xiang, Y. (2005) Plasticity in Cu thin films: an experimental investigation of the effect of microstructure, Ph.D. thesis, Harvard University.

Xiang, Y., Chen, X., and Vlassak, J. J. (2005) The plane-strain bulge test for thin films, *J. Mater. Res.,* 20, pp. 2360–2370.

Yonezu, A., Kuwahara, Y., Yoneda, K., Hirakata, H., and Minoshima, K. (2009) Estimation of the anisotropic plastic property using single spherical indentation—an FEM study, *Comp. Mater. Sci.,* 47, pp. 611–619.

Yu, D. Y. W., and Spaepen, F. (2004) The yield strength of thin copper films on Kapton, *J. Appl. Phys.,* 95, pp. 2991–2997.

Zhao, M., Chen, X., Xiang, Y., Vlassak, J. J., Lee, D., Ogasawara, N., Chiba, N., and Gan, Y. X. (2007) Measuring elastoplastic properties of thin films on an elastic substrate using sharp indentation, *Acta Mater.,* 55, pp. 6260–6274.

Zhang, T., Su, Y., Qian, C. F., Zhao, M. H., and Chen, L. Q. (2000) Microbridge testing of silicon nitride thin films deposited on silicon wafers, *Acta Mater.,* 48, pp. 2843–2857.

Zhou, Q., Itoh, G., and Yamashita, T. (1999) Further study on the effects of specimen thickness and grain size on the creep behavior of aluminum alloy foils. *Mater. Trans., JIM,* 40, pp. 443–446.

Chapter 4

GROWTH OF INTERFACE CRACK

4.1 INTERFACE FRACTURE TOUGHNESS

4.1.1 Measurement Methods

Interface fracture toughness of bulk bi-materials has been evaluated on the basis of the fracture mechanics concept (described in Chapter 2), where the conventional experimental methods have been established at the practical level in laboratories [e.g., Rice, 1988; Hutchinson and Suo, 1992; Evans and Dalgleish, 1992]. But evaluating the interface toughness between submicron- or nanometer-thick films and between the films and the substrate presents difficulties because the fabricating and loading such small specimens is restrictive. Although many testing techniques have been proposed [e.g., Evans et al., 1999; Volinsky et al., 2002; Nied, 2003; Thouless, 1994; Lane, 2003], we do not have universal method yet. This section briefly reviews the proposed methods. Table 4.1 summarizes the advantages and disadvantages of each from the standpoint of the applicability of fracture mechanics and the feasibility of sample preparation and handling.

Fracture Nanomechanics
Edited by Takayuki Kitamura, Hiroyuki Hirakata, Takashi Sumigawa and Takahiro Shimada

www.panstanford.com

4.1.1.1 Indentation method

Table 4.1 Methods for evaluating interface fracture toughness of thin films

Method	Advantage	Disadvantage	Reference
Indentation	Easy sample preparation	Only for weak interface Need of indented volume, delamination area, and residual stress measurements	Marshall and Evans, 1984; Rossington et al., 1984
Superlayer indentation	Easy sample preparation Applicable for strong interface	Need of indented volume, delamination area, and residual stress measurements	Kriese et al., 1999a, 1999b
Line scratch test	Direct measurement of lateral load	Specimen preparation by lithography	De Boer et al., 1997
Projection test	Direct measurement of delamination load	Specimen preparation by etching	Kamiya et al., 2002
Superlayer test	No need for mechanical loading (gripping of sample)	Specimen preparation by lithography Need of residual stress measurements	Bagchi et al., 1994; Bagchi and Evans, 1996
Four-point bend test	Applicability of fracture mechanics No need for crack length measurement Constraint of deformation and fracture of thin film Insensitive to residual stress	Need of substrate bonding	Charalambides et al., 1989; Dauskardt et al., 1998
Double cantilever beam/ Compact tension test	Applicability of fracture mechanics Constraint of deformation and fracture of thin film Insensitive to residual stress	Need of substrate bonding	
Cantilever method	Applicability of fracture mechanics Constraint of deformation and fracture of thin film Insensitive to residual stress	Need of substrate bonding	Hirakata et al., 2005
Bulge test	Easy control of load	Specimen preparation by lithography Only for strong films Large deformation of films	Hohfelder et al., 1997; Jensen, 1991
Peel test		Only for strong films Large deformation of films	

With the indentation method [Marshall and Evans, 1984; Rossington et al., 1984], an indenter with an angular tip (e.g., conical indenter) is pressed into a thin film on a substrate, as shown in Fig. 4.1, and an impression is formed by the plastic deformation near the tip contact. The displaced volume causes compressive stress in the thin film, and the stored elastic strain energy is released to induce a radial interface crack. The crack is then arrested at an equilibrium length where the crack driving force is equal to the interface fracture toughness. The measurements of plastically displaced volume V0 and delamination radius (crack length) a are used to calculate the critical energy release rate GC at the arresting point. The energy release rate G of the indentation-induced delamination [Marshall and Evans, 1984] has the following form.

$$G = \frac{h\sigma_V^2(1-\nu_f^2)}{2E_f} + \frac{(1-\alpha_{BE})\,h\sigma_R^2(1-\nu_f)}{E_f} + \frac{(1-\alpha_{BE})\,h(\sigma_V-\sigma_B)^2(1-\nu_f)}{E_f} \tag{4.1}$$

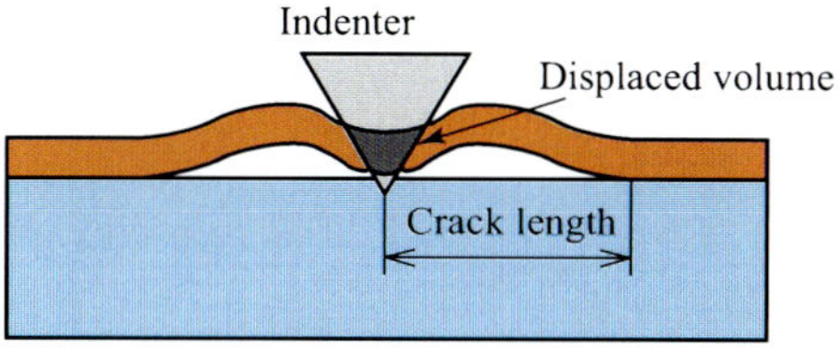

Figure 4.1 Schematic cross-section of the indentation test.

Here, σ_R is the residual stress of the thin film introduced during the deposition process, σ_V is the film stress caused by the displaced volume V_0, σ_B is the buckling stress of the thin film, α_{BE} represents the slope of the buckling load versus the edge displacement [Volinsky et al., 2002], h is the film thickness, and E_f and ν_f are the Young's modulus and Poisson's ratio of film. Since σ_V and σ_B are in inverse proportion to a^2, the energy release rate G is a decreasing function of a. By substituting the measured crack length a into Eq. (4.1), the critical energy release rate G_C or the interface toughness at the arresting point is evaluated. It should be noted that the fracture toughness at the arresting point may differ from the conventional definition of fracture toughness for the initiation of crack propagation because the plastic deformation history ahead of the crack tip between them is different.

We can use the indentation method to evaluate interface toughness without having to do any special preparation of the specimens. However, the precise evaluation of the interface toughness may be difficult since a complex stress field is induced by the large plastic deformation and fracture of the thin film and substrate near the chip contact region.

4.1.1.2 Superlayer indentation method

For a low stiffness film, indentation alone cannot supply enough driving force to delaminate the film from the substrate. A modified technique to overcome the issue is to add a highly stressed and high stiffness superlayer on top of the thin film, as shown in Fig. 4.2 [Kriese et al., 1999a, 1999b; Gerberich et al., 1999; Tymiak et al., 1999]. The superlayer adds a driving force to delamination. Because the basic concept is similar to the one in the indentation method, the energy release rate is analyzed for multi-layered films in a similar way [Kriese et al., 1999a].

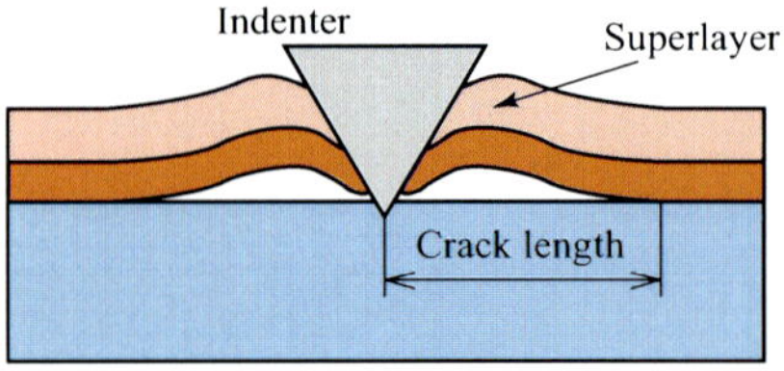

Figure 4.2 Schematic cross-section of the superlayer indentation test.

4.1.1.3 Line scratch test

In a conventional scratch test, a thin film is delaminated from a substrate by scratching the film surface with a diamond stylus. A normal ramp load is applied to the film by the laterally moving indenter tip until delamination takes place, and the critical load gives a measure of the interface strength. Although the testing procedure is simple, it is difficult to extract the pure interface toughness because film fracture is inevitable during the test. Attempts at quantitatively evaluating interface toughness and delamination stress have been conducted by analyzing complex deformation in the scratching processes [Venkataraman et al., 1996; Sekiguchi and Koike, 2008], but the results have not been completely clear.

Using a pre-cracked line structure to conduct a line scratch test has been proposed [De Boer et al., 1997]. After the line structure is fabricated by a lithography technique, the lateral force is directly applied to the line edge by an indenter wedge, as shown in Fig. 4.3. A carbon release layer is partly inserted between the line and the substrate in the preparation process to introduce an interface pre-crack by a low load.

Because the contact region is away from the crack tip, the complex plastic deformation hardly affects the stress field near the crack tip. However, this method of specimen preparation is not easy, which offsets the advantage of simple evaluation in the original scratch method. In addition, the shape of the pre-crack tip is not ideally sharp due to the carbon release layer, and this might affect the accuracy of the obtained fracture toughness.

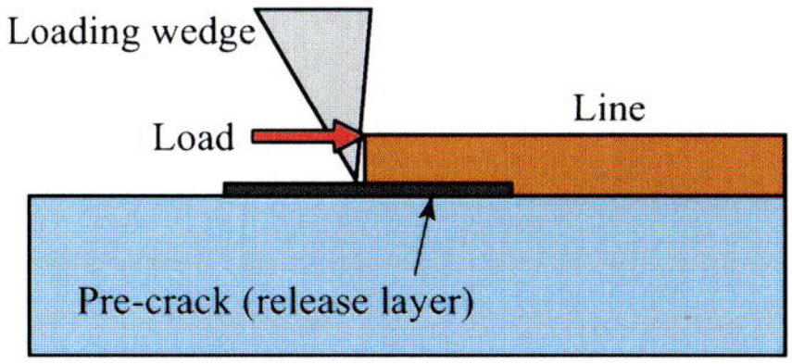

Figure 4.3 Schematic illustration of the line scratch test.

4.1.1.4 Projection test

For a projection test, a specimen with a film projection (eaves) is fabricated by an etching process, and the delamination load is applied by pushing up the projection with an indenter tip, as shown in Fig. 4.4 [Kamiya et al., 2002]. Preliminary loading introduces an interface crack, which grows radially from the edge near the loading point. The length of the crack is measured to evaluate the interface fracture toughness at the crack arresting point. The difficulty lies in precisely measuring crack size and shape.

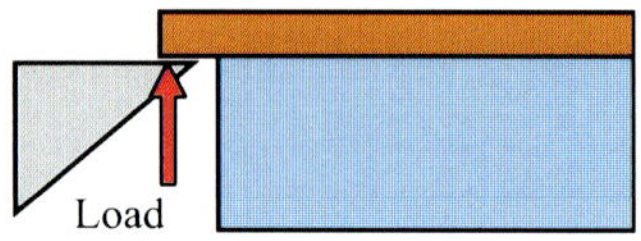

Figure 4.4 Schematic illustration of the projection test.

4.1.1.5 Superlayer test

During the film fabrication process, residual stress in a thin film sometimes brings about delamination. In the superlayer test [Bagchi et al., 1994; Bagchi and Evans, 1996], which takes advantage of this phenomenon, a highly stressed additional film (superlayer) is deposited on the top of a specimen to increase the driving force of delamination. Figure 4.5 shows an example of the interface toughness evaluation using this concept, which includes three steps. First, a carbon release layer that acts as a pre-crack is deposited on a substrate (Step 1). A thin film of interest and a superlayer are then deposited and patterned to form a strip (Step 2). Finally, the bi-layer is cut by etching or ion beam milling (Step 3). Since the applied stress intensity factor K (or the energy release rate G) depends on the thickness of the superlayer, the experiments are repeated on samples with different thicknesses of superlayers. If the crack driving force (stress intensity factor or energy release rate) due to the residual stresses in the bi-layer exceeds the interface fracture toughness K_C (or G_C), the delamination crack begins to propagate. Thus, the magnitude of K_C (or G_C) is determined by the critical thickness and the toughness is equivalent to that for the initiation of crack propagation. In this method, because no external force is needed, complex plastic deformation due to the chip contact is avoided. However, the method requires lithography techniques, and several specimens with different thicknesses of superlayer have to be prepared. The difficulty lies in identifying the precise residual stresses of different specimens. Moreover, the blunt crack tip (comparing with the thickness of thin film) may affect the magnitude of toughness.

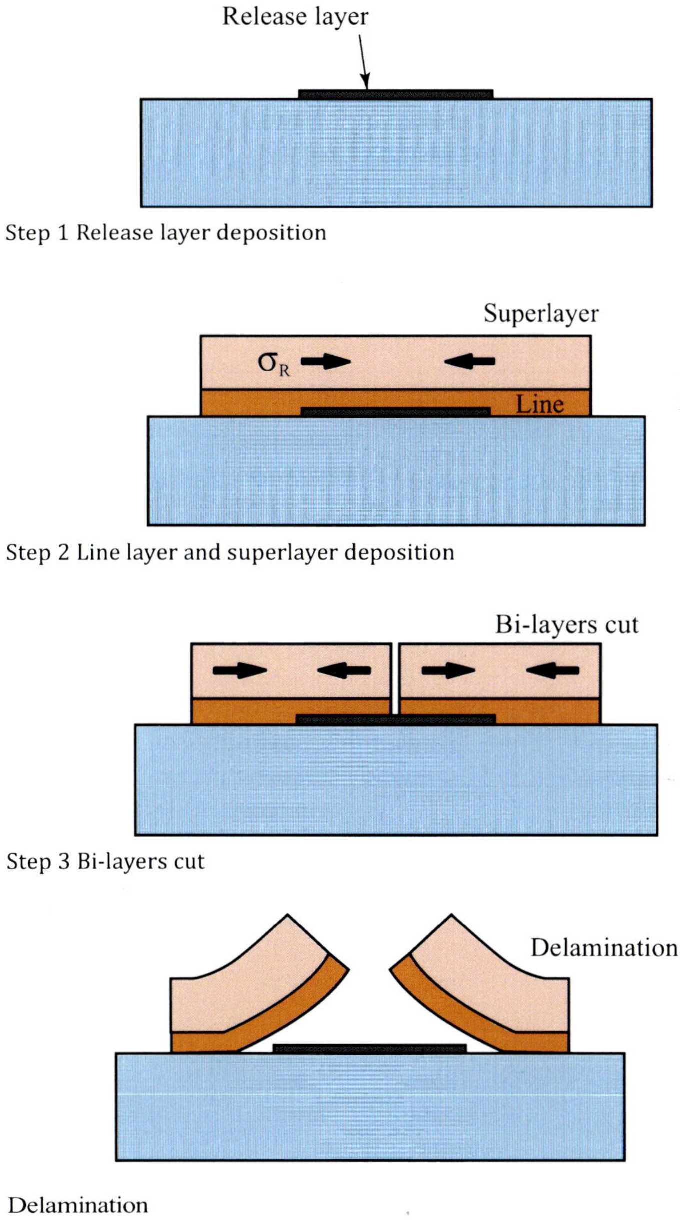

Figure 4.5 Procedure of the superlayer test.

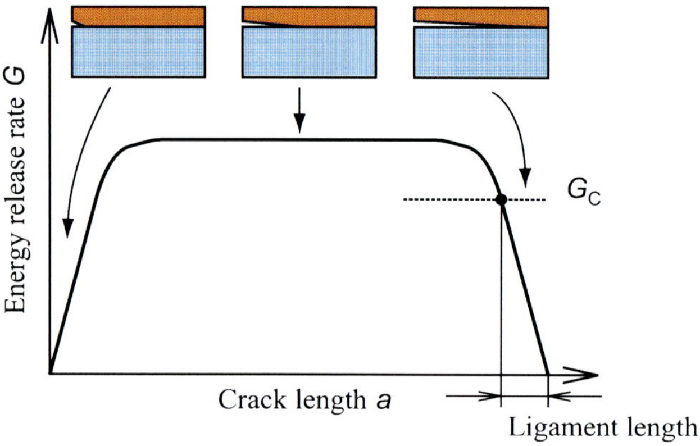

Figure 4.6 Schematic diagram of the three stages of debonding along an interface between a residually stressed film and a substrate.

Figure 4.6 shows a schematic relationship between the crack length a and the energy release rate G of an interface crack of a residually stressed thin film on a substrate. When the crack tip locates away from the edge, the energy release rate is almost constant (steady state). As the crack tip approaches the edge, the G value decreases. In a proposed method [He et al., 1997; Zhuk et al., 1998], the interface fracture toughness G_C at the crack arrest can be determined by the residual ligament length, as schematically shown in Fig. 4.6.

In a similar method, the thermal stress caused by the mismatch of coefficient of thermal expansion can be used as a driving force for delamination [Hay et al., 2001; Buchwalter, 2001]. In this method, a polymeric superlayer (e.g., epoxy) is deposited on top of a thin film/substrate structure, and then temperature is reduced until the delamination occurs due to the thermal stress.

4.1.1.6 Four-point bend test

Conventional fracture mechanics testing methods for bulk materials have been applied to testing thin film/substrate structures with appropriate modifications. The four-point bend method [Charalambides et al., 1989; Dauskardt et al., 1998; Dauskardt and Lane, 2000] is one of the most popular methods. A symmetric four-point bending load is applied to a sandwich specimen with two substrates, as shown in Fig. 4.7. The stiff substrates effectively constrain plastic deformation and fracture of the thin film.

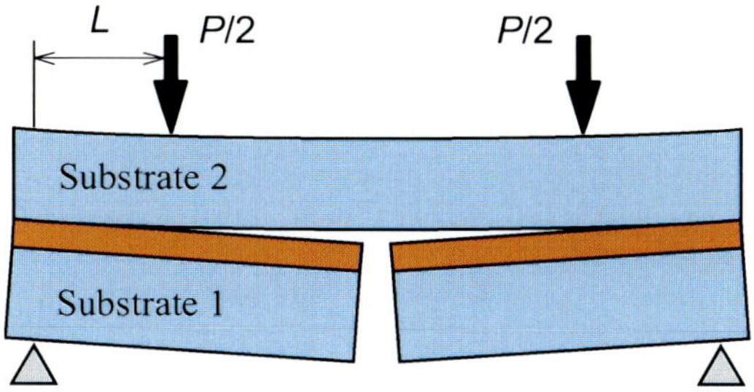

Figure 4.7 Schematic illustration of the four-point bend test.

The remarkable merit of the method is that the energy release rate G does not depend on the crack length a. Thus, we can evaluate the interface fracture toughness without measuring the crack length.

The energy release rate G of the specimen configuration is expressed as follows.

$$G = \frac{P^2L^2\left(1-\nu_2^2\right)}{8E_2b^2}\left(\frac{1}{I_2}-\frac{\lambda}{I_c}\right)$$
$$\lambda = \frac{E_2\left(1-\nu_1^2\right)}{E_1\left(1-\nu_2^2\right)}, I_2 = \frac{h_2^3}{12}, I_c = \frac{h_1^3}{12}+\lambda\frac{h_2^3}{12}+\lambda\frac{h_1h_2\left(h_1+h_2\right)^2}{4\left(h_1+\lambda h_2\right)} \tag{4.2}$$

Here, P is the load, L is the distance between two loading points, b is the width of the specimen, and E_i, ν_i, and h_i (i = 1, 2) are Young's modulus, Poisson's ratio, and the thickness of the substrates. The subscripts 1 and 2 indicate the lower and upper substrates, respectively.

Since the film deposited on a substrate (Substrate 2 in Fig. 4.7) must be bonded to another substrate (Substrate 1 in Fig. 4.7) by epoxy adhesives or by diffusion bonding of metal layers to make the sandwich structure, the interface of interest has to be weaker than the adhesive layer. Thus, this method cannot be applied for structures with a strong interface.

4.1.1.7 Double cantilever beam/Compact tension test

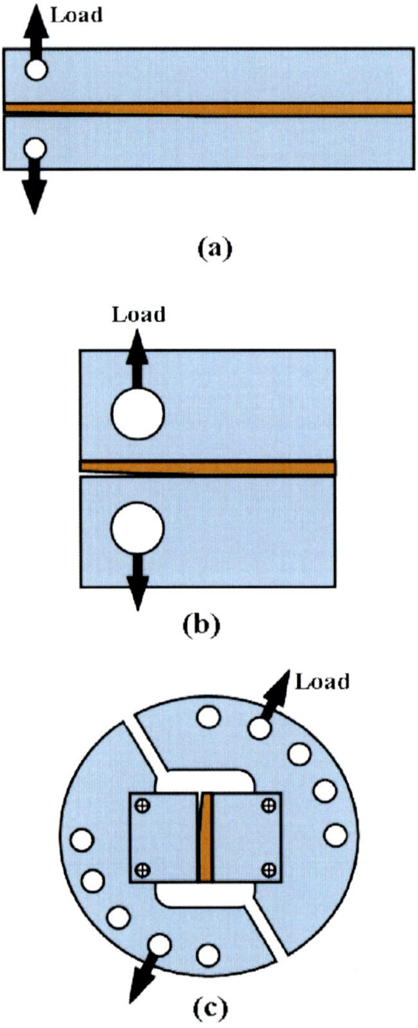

Figure 4.8 S chematic illustrations of (a) the double cantilever beam test, (b) the compact tension test, and (c) the Brazil-nuts tests.

Other popular testing methods using a sandwiched specimen following the macroscopic fracture mechanics methodology such as the double cantilever beam testing (DCB, Fig. 4.8a), the compact tension (CT, Fig. 4.8b), and the Brazil-nuts (Fig. 4.8c) [Wang and Suo, 1990] can also be applied to thin film/substrate structures. The characteristics of these methods are similar to the four-point bend method mentioned in the previous section, although these methods require measuring crack length to estimate the crack driving force.

4.1.1.8 Cantilever method

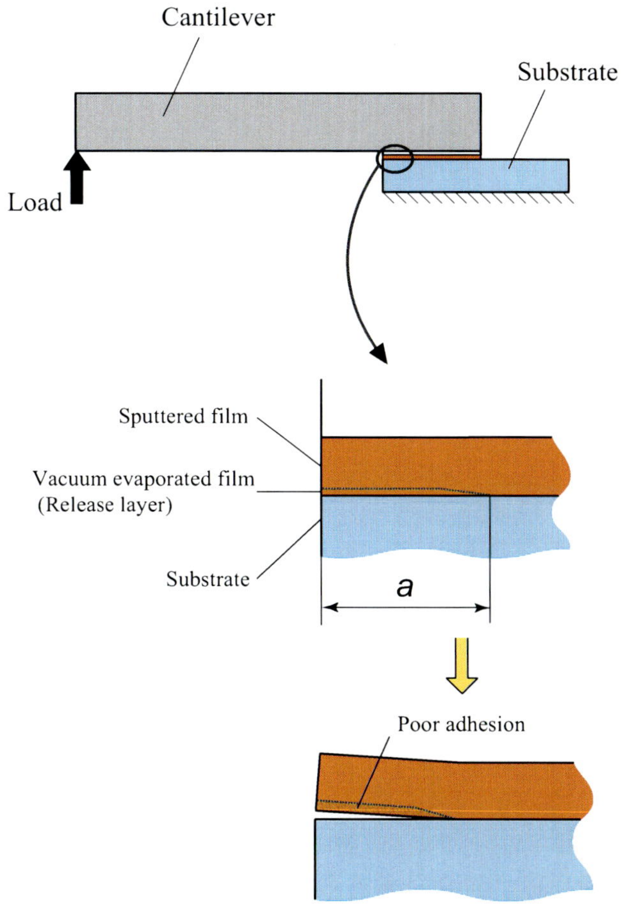

Figure 4.9 Schematic illustration of the cantilever method. A pre-crack is introduced along a thin film/substrate interface by using the difference in interface strength by the deposition method.

A cantilever beam for applying delamination load is bonded on a thin film/substrate sample as shown in Fig. 4.9 [Hirakata et al., 2005]. In this method, a film/substrate interface consisting of strong and weak parts is produced by using two different deposition methods. After an ultra-thin film (release layer) that has poor adhesion to its substrate is partly formed on the substrate by a deposition method such as vacuum evaporation, the identical material is deposited by a different method, such as sputtering, which produces a stronger interface than the first layer. The release layer debonds from the substrate at low load, which introduces a sharp pre-crack along the interface. The boundary between the release layer and the thin film (the dashed line in Fig. 4.9) is stronger than the bi-material interface because the deposited materials are identical and no modulus mismatch exists. The advantage is that the initial crack length a is known and controllable. This is applicable to the evaluation of a strong interface. Of course, this idea can be extended to the other methods discussed before.

4.1.1.9 Bulge test

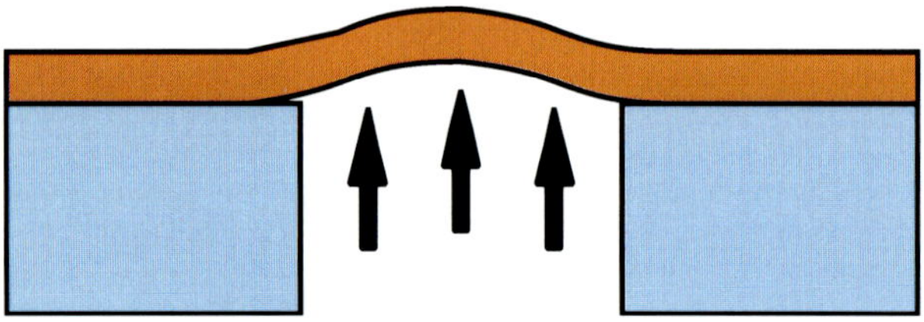

Figure 4.10 Schematic cross-section of the bulge test.

A free-standing thin film at a window of substrate is formed by MEMS fabrication using substrate etching, and the window is pressurized from the bottom by means of a gas to apply the driving force for delamination, as shown in Fig. 4.10 [Hohfelder et al., 1997; Jensen, 1991]. The method requires complicated sample preparation and cannot be applied to weak thin films because of the inevitable film fracture and large plastic deformation.

4.1.1.10 Peel test

In the peel test (Fig. 4.11), the edge of a film is chucked and a peeling force is applied to the film. A crack tip or a point of separation steadily propagates along the interface. An additional layer is required to reinforce a weak film to prevent film fracture. An improvement of this method is the multi-stages test, where the specimen is put on a movable stage and the load is applied to the interface by attaching dead weight to the film through a pulley [Omiya et al., 2004]. This dead weight works as a shear force at the peeling tip by the proposed lateral loading while the peeling angle can be readily controlled.

Most of the methods shown in this section can be applied only to layered materials that extend two dimensionally, i.e., thin films, and most reported examples of evaluation are for micron- or submicron-meter-thick films. Few techniques have been developed to evaluate the interface fracture toughness of three-dimensionally small components on substrates especially in the nano-scale. This is due to the difficulty in loading to such small-components and in observing the interface crack propagation.

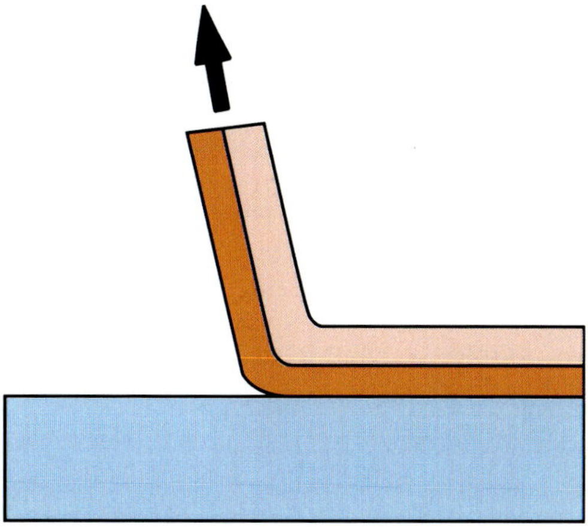

Figure 4.11 Schematic illustration of the peel test.

4.1.2 Applicability of Fracture Mechanics Concept to Delamination of Submicron Film

This section describes evaluating the fracture toughness of an interface between a 200 nm-thick Cu film and a Si substrate by the four-point bend method, adopting a modified specimen shown in Fig. 4.12 [Hirakata et al., 2006; Truong et al., 2006]. Because the standard four-point bend specimen has two symmetrical cracks, the same singular stress field theoretically takes place near both crack tips under an applied load. However, in reality, it is difficult to control the equal growth of the cracks because a slight unbalance in the loading brings about asymmetrical stress intensities. Using the modified specimen, which has only one interface crack, enables us to control the stable crack growth.

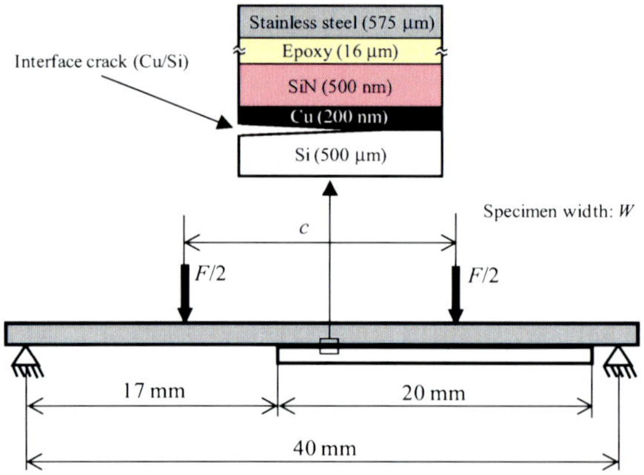

Figure 4.12 Modified four-point bend specimen for crack propagation experiment along an interface between a thin film and a substrate. Reprinted from Hirakata, H., et al. (2007) "Role of plasticity on interface crack initiation from a free edge and propagation in a nano-component," *Int. J. Fract.*, 145, pp. 261–271, with permission from Springer.

Figure 4.13 shows the relationship between the applied load and the displacement at the loading point obtained by experiment. The load becomes almost constant or a plateau region appears during stable crack propagation. This is because the stress intensity factor (or the energy release rate) barely depends on the crack length in the four-point bending method under a constant load, similar to the original four-point bend described in Section 4.1.1.6 [Charalambides et al., 1989; Dauskardt et al., 1998]. The critical load for interface crack propagation can be easily determined as the load at the plateau region.

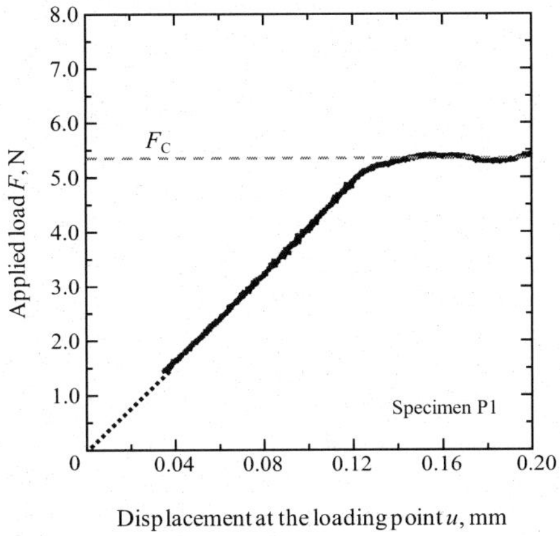

Figure 4.13 Relationship between load and displacement in a modified four-point bend experiment. The plateau region indicates steady-state crack propagation. Reprinted from Hirakata, H., et al. (2007) "Role of plasticity on interface crack initiation from a free edge and propagation in a nano-component," *Int. J. Fract.*, 145, pp. 261–271, with permission from Springer.

Figure 4.14 shows the elastic stress distributions along the Cu/Si interface ($\theta = 0$) at crack propagation. The singular stress field along the interface near the crack tip is characterized by complex stress intensity factors [Erdogan, 1963], K_1 and K_2, defined as

$$\sigma_{\theta\theta} + i\tau_{r\theta} = \frac{K_1 + iK_2}{\sqrt{2\pi r}}\left(\frac{r}{l}\right)^{i\varepsilon} \qquad \text{same as (2.28)}$$

where ε is the bi-material constant determined by the elastic constants of the materials, r is the distance from the crack tip, and l is the characteristic length (see Chapter 2) [Rice, 1988; Ikeda et al., 1998]. The figure shows that the shear stress is dominant in comparison with the normal stress; the phase angle $\Phi = \tan^{-1}(K_2/K_1)$ in this case is about −61 deg at a characteristic length of $l = 10$ nm. The critical total stress intensity factor,

$$K_i = \sqrt{K_1^2 + K_2^2} \tag{4.3}$$

is evaluated as K_{iC} = 0.74 ± 0.02 MPam$^{0.5}$ from the plateau load. The corresponding critical energy release rate is G_C = 3.6 ± 0.2 J/m^2.

The dominant region of the singular stress field is on the order of 10 nm for normal stress and 100 nm for shear stress. If the plastic region is much smaller than the elastic singular stress region, the small-scale yielding (SSY) condition is satisfied. In this case, the singular field governs crack propagation and hence the fracture toughness can be evaluated by K_i or G, as discussed above. However, the normal and shear stresses in the singular region of 10 nm are on the order of 1~10 GPa analyzed by the linear-elastic theory shown in Fig. 4.14, and hence large plastic deformation might take place near the crack tip, the large-scale yielding (LSY) condition.

Since the size of the singular stress field is generally small in thin films, the interface crack between metallic thin films is likely not to satisfy the SSY condition in general. In the LSY condition, K_i does not characterize the intensity of the singular stress field anymore. A large amount of energy dissipates because of the plastic deformation, and thus G loses its physical meaning as the energy required for the interface separation (the true work of adhesion). In this case, the K_{iC} or G_C value that is obtained by assuming the linear elasticity is not the inherent material property but instead depends on factors such as film thickness.

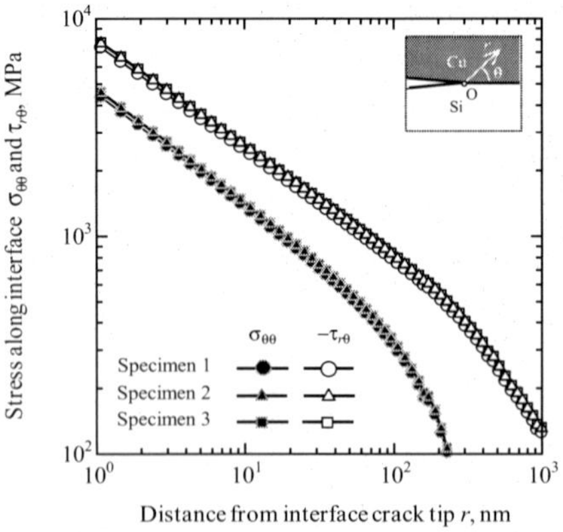

Figure 4.14 Elastic stress distributions near the interface edge between Si and Cu at crack propagation.

Using the bend method of cantilever shown in Chapter 3 (Section 3.4.4.1), we can evaluate the plasticity of the Cu layer. The Cu layer obeys the Ramberg-Osgood law,

$$\bar{\varepsilon} = \frac{\bar{\sigma}}{E} + B\bar{\sigma}^n \tag{4.4}$$

where $\bar{\sigma}$ and $\bar{\varepsilon}$ are the von Mises stress and strain, B and n are the material constants characterizing the plastic property; B = 0.10 GPa^{-n} and n = 3.2 [Takahashi et al., 2007] (It should be noted that the magnitude of n is different from the one in Chapter 3 because Eq. (4.4) is different from Eq. (3.9')). We can discuss the applicability of the elasto-plastic fracture mechanics on the basis of the nonlinear constitutive equation.

Figure 4.15 shows the distributions of normal and shear stresses along the Cu/Si interface considering the plasticity of Cu. The plastic region is much larger than the elastic region, proving that the SSY condition is not satisfied. The singular stress field near the interface crack tip is approximately expressed by the following HRR-like equation (see Section 2.4.3),

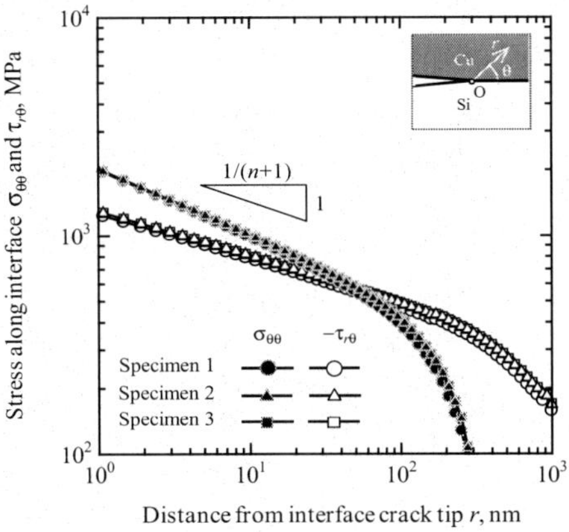

Figure 4.15 Elasto-plastic stress distributions near the interface edge between Si and Cu at crack propagation. The Cu layer is assumed to obey the Ramberg-Osgood power law of plasticity.

$$\sigma_{ij} = \left(\frac{J}{I_n B r} \right)^{\frac{1}{n+1}} \tilde{\sigma}_{ij}(\theta) \tag{4.5}$$

where J is the J-integral wich is a path-independent line integral around the crack tip and characterizes the intensity of the elasto-plastic singular stress field [Rice, 1968], and $\tilde{\sigma}_{ij}(\theta)$ is the nondimensional function of θ. The interface toughness is evaluated by using the J-integral as $J_C = 3.6 \pm 0.2$ J/m^2. The ratio of normal and shear stress depends on r in the singular region, and the $\tan^{-1}(\tau_{r\theta}/\sigma_{\theta\theta})$ at $r = 10$ nm is about -39 deg. The region size of the singular stress field is on the order of 10 nm for normal stress and 100 nm for shear stress. The applicability of fracture mechanics for such small singular stress fields is discussed in detail in Chapter 5.

4.2 STABLE CRACK GROWTH

4.2.1 Fatigue

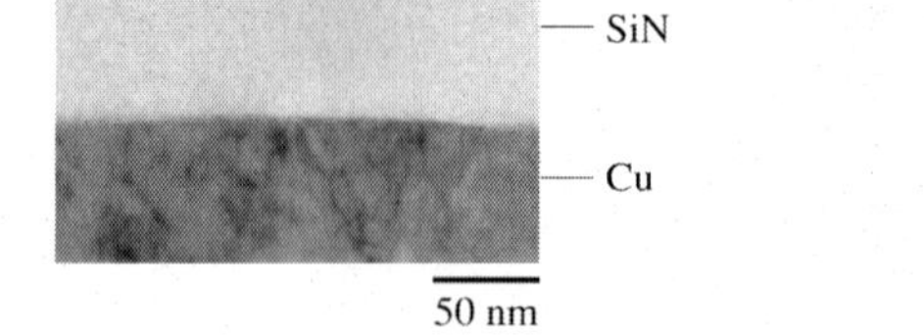

Figure 4.16 Schematic illustration of tested multi layered thin films on a Si substrate and a TEM micrograph of the region near the interface between Cu and SiN. Reprinted from Hirakata, H., Kitazawa, M., and Kitamura, T. (2006) "Fatigue crack growth along interface between metal and ceramics submicron-thick films in inert environment," *Acta Mater.*, 54, pp. 89–97, Copyright (2006), with permission from Elsevier.

Fatigue crack growth is caused by cyclic plastic deformation of metal near the crack tip in general. Many researchers have investigated it along interfaces in a bulk bi-material [e.g., McNaney et al., 1996; Shaw et al., 1994; Reimanis et al., 1990; Okazaki et al., 2001; Gao et al., 2003; Mutoh et al., 2003]. The interface cracking between aluminum (Al) and alumina (Al_2O_3) under fatigue [McNaney et al., 1996] suggests the mechanism of crack growth; the striations observed on the fracture surface of the Al side signify that the blunting and resharpening of the crack tip takes place every cycle. The conventional fracture mechanics of fatigue crack growth in metal can thus be applied to the cracks along the interface of dissimilar bulk materials. In fact, the crack growth rate in dissimilar bulk materials is uniquely characterized by the stress intensity factor range ΔK or the energy release rate range ΔG [McNaney et al., 1996; Mutoh et al., 2003].

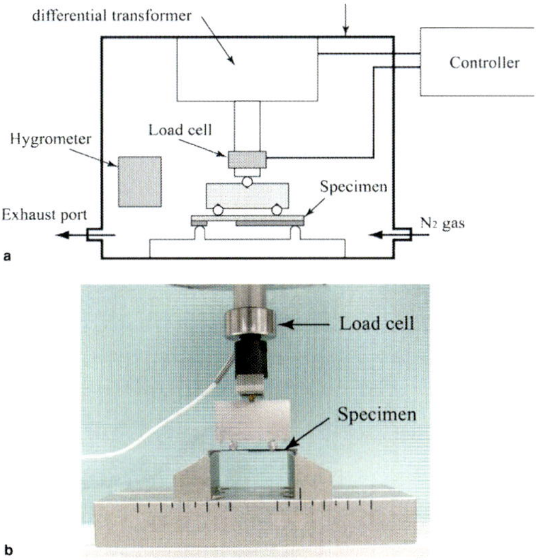

Figure 4.17 Testing apparatus for the fatigue experiments: (a) schematic of testing system; and (b) photograph of specimen and loading jig. A hermetic chamber is attached to the testing apparatus and dry nitrogen (N_2) gas is pumped into the chamber to realize an inert environment. Reprinted from Hirakata, H., Kitazawa, M., and Kitamura, T. (2006) "Fatigue crack growth along interface between metal and ceramics submicron-thick films in inert environment," *Acta Mater.*, 54, pp. 89–97, Copyright (2006), with permission from Elsevier.

On the other hand, for submicron-thick films, few studies [Hirakata et al., 2006] have been conducted for fatigue crack growth. Since the dislocation motion is restricted in a thin layer of metal film [Schwaiger and Kraft, 2003; Schwaiger et al., 2003] and the plastic deformation of the film is constrained by its stiff substrates [Reimanis et al., 1990; McNaney et al., 1996; Kruzic et al., 2004], fatigue crack growth behavior along the interface between submicron-thick films differs from that in bulk materials. In this section, we discuss fatigue crack growth behavior along the interface between submicron-thick sputtered Cu and PECVD SiN films, as shown in Fig. 4.16. Modified four-point bend specimens similar to those in the previous section are used for the experiments. In order to evaluate fatigue behavior caused by purely mechanical loading, a hermetic chamber is attached to the testing section of the loading apparatus, and dry nitrogen (N_2) gas is pumped into the chamber during the tests to reduce the relative humidity to 3.5% ± 1% (Fig. 4.17).

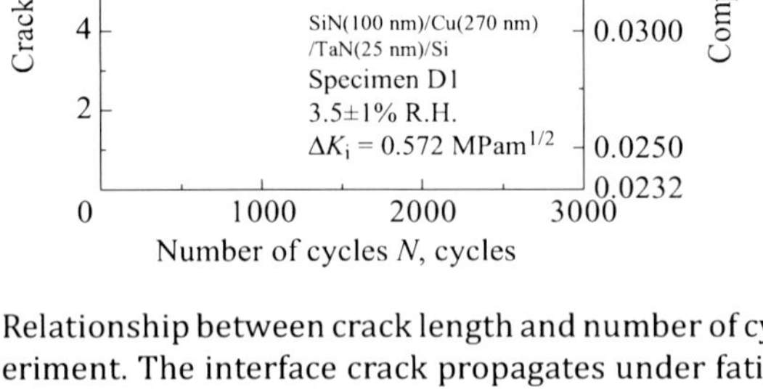

Figure 4.18 Relationship between crack length and number of cycles during a fatigue experiment. The interface crack propagates under fatigue loading even in the inert environment. Reprinted from Hirakata, H., Kitazawa, M., and Kitamura, T. (2006) "Fatigue crack growth along interface between metal and ceramics submicron-thick films in inert environment," *Acta Mater.*, 54, pp. 89–97, Copyright (2006), with permission from Elsevier.

Figure 4.18 shows the crack growth curve for a specimen, where the measured compliance is indicated on the right side as a reference. The crack begins to propagate along the Cu/SiN interface just after the test is started, and the crack growth rate is almost constant. This signifies the purely mechanical crack growth in fatigue between the thin films. The crack growth then slows down and stops. This is because the crack approaches the inner loading points and the magnitude of stress intensity factor range ΔK_i (see Eq. (4.3)) decreases rapidly.

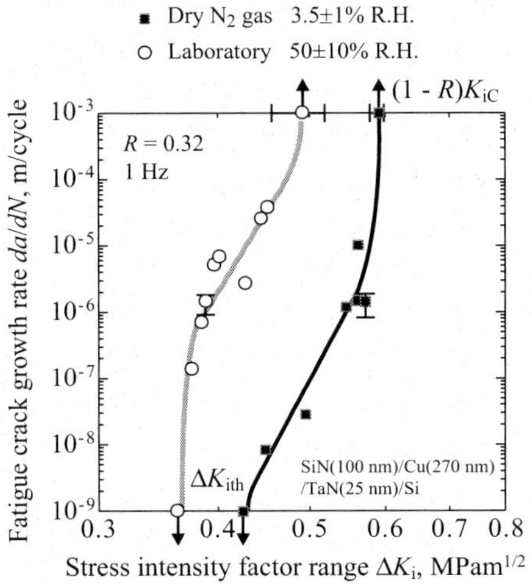

Figure 4.19 Relationship between fatigue crack growth rate and stress intensity factor range in inert and laboratory environments. The relationship is represented by the sigmoidal line consisted of the three stages: threshold, stable growth, and critical growth similar to typical fatigue behavior in bulk. Reprinted from Hirakata, H., Kitazawa, M., and Kitamura, T. (2006) "Fatigue crack growth along interface between metal and ceramics submicron-thick films in inert environment," *Acta Mater.*, 54, pp. 89–97, Copyright (2006), with permission from Elsevier.

Figure 4.19 shows the relationship between crack growth rate, *da/dN*, and stress intensity factor range, ΔK_i, in the inert gas. There exists a good correlation between *da/dN* and ΔK_i, indicating that the stress intensity near the tip governs the crack growth and the sigmoidal line consists of the three stages: near threshold, stable growth, and critical growth similar to typical fatigue behavior in a bulk. The downward arrow indicates the threshold, ΔK_{ith}, where the crack does not propagate in 2×10^5 cycles. The upward arrow indicates the equivalent interface fracture toughness, $(1 - R)K_{iC}$, where R is the ratio K_{imin}/K_{imax}.

In general, the ratio of threshold to fracture toughness, $\Delta K_{ith}/K_{iC}$, in brittle metals is of high magnitude, while the ratio in ductile materials is relatively low. The value of $\Delta K_{ith}/K_{iC}$ in the fatigue of the Cu/SiN interface is about 0.49, which is about five times higher than that of bulk Cu (≈ 0.1 [Fleck et al., 1994]) and interfaces in metal/ceramics bulk bi-materials (e.g., Al and Al_2O_3: ≈ 0.1 [Fleck et al., 1994; McNaney et al., 1996]). Thus, crack growth along the interface between the Cu/SiN films is very brittle. As shown in Chapter 3, a thinner film has a higher yield stress, which causes crack propagation with small plastic deformation. Thus, the deformation of films strictly constrained by the hard substrate characterizes the brittle growth behavior.

The experimental result in the laboratory environment (50 ± 10% R.H.) is also plotted in the figure, allowing us to examine the influence of atmosphere. It is clear that the crack growth is enhanced by the presence of moisture. The effect of moisture on crack growth is more noticeable in the stable growth region than in the threshold and the fracture toughness regions. It is well known that a crack subcritically grows under a static load in moist air (static fatigue) [Dauskardt et al., 1998; Lane et al., 2001; Hughey et al., 2004; Xu et al., 1999]. The major mechanism of static fatigue is stress corrosion, in which a chemical reaction near the crack tip plays an important role in crack growth [Wiederhorn, 1967; Magnin 1995]. Although the acceleration of fatigue crack growth shown in Fig. 4.19 is basically attributed to stress corrosion, chemical contributions are quantitatively different in static fatigue and in cyclic fatigue [Rolfe and Barsom, 1977; Minoshima and Komai, 1988].

Fatigue behavior in thin films plays an important role in fatigue crack growth along a thin film interface. Although several investigations have been conducted into fatigue in thin films [Schwaiger and Kraft, 2003; Schwaiger et al., 2003; Zhang et al.,

2006], elucidating the detail mechanics including the applicability of fracture mechanics remains for future work.

In thinner or nano-thick films, fatigue is expected to be more restricted since the dislocation motion in the films is severely constrained by the substrate and fewer dislocation sources exist in the films. In such cases, it is not known whether the interface crack grows as the result of fatigue. Moreover, interface crack initiation and propagation due to fatigue in three-dimensionally small components have not been elucidated, and the methods for exploring it are needed. Since the size of the singular stress field becomes smaller, as pointed out in the previous section, studies on the applicability of fracture mechanics for such nano-components remain to be explored in the future.

4.2.2 Creep

Creep crack growth behavior of bulk metals at elevated temperatures has been experimentally investigated on the basis of the nonlinear fracture mechanics [Landes and Begley, 1976; Taira et al., 1979]. Creep crack growth is governed by the singular field of stress or strain rate (HRR field) near the crack tip, and crack growth rate is correlated well with the creep J-integral J^* that characterizes the intensity of the stress field and strain rate field under large-scale creep (LSC) conditions. On the other hand, Biner [1997a, 1997b] numerically revealed that the stress field ahead of a stationary interface crack in elastic-creeping bi-materials resembles the HRR filed (see Chapter 2). However, it is not clear whether the stress fields actually govern the crack growth rate even in bulk materials because few experimental studies [Al-Khanbashi et al., 2003] have been conducted on the creep crack growth along an interface. In particular, the applicability of fracture mechanics to submicron- or nanometer-thick films has not been studied well.

As schematically shown in Fig. 4.20, the creep region at a crack tip is small at the loading instant, and the elastic stress singularity is dominant (SSC). The creep region expands with the passage of time, and it becomes steady state (LSC), as explained in Section 2.3.2.2. In a submicron-thick film that is usually fixed on a substrate or sandwiched by two substrates, large creep deformation hardly expands, because of the constraint. Therefore, it is expected that the interface crack propagates before the creep region spreads large enough or the stress field and strain rate field reach the steady state (LSC).

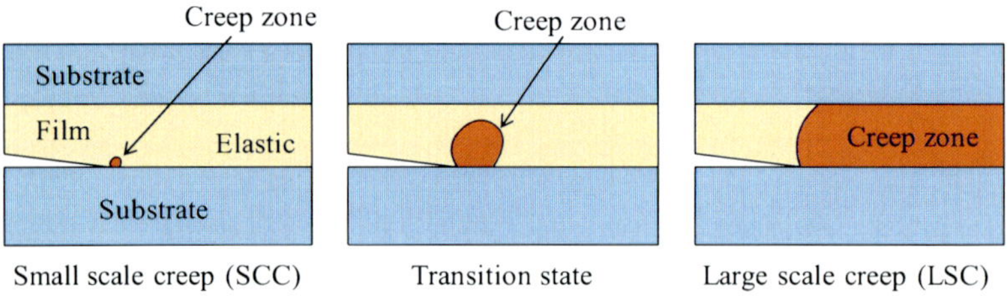

Figure 4.20 Schematic illustration of the transition of creep zone ahead of an interface crack between a thin film and a substrate.

In order to clarify the creep crack growth behavior of submicron-thick films on substrates, we prepare a cantilever specimen, a 300 nm-thick epoxy film and a 1 mm-thick soda-lime glass on a Si substrate, as shown in Fig. 4.21. The creep property (Norton's power law) of the epoxy film is

$$\dot{\bar{\varepsilon}} = 2.89 \times 10^{-16} \bar{\sigma}^{4.77} \tag{4.6}$$

where $\bar{\sigma}$ and $\dot{\bar{\varepsilon}}$ are the von Mises equivalent stress and the von Mises equivalent strain rate. A constant load is applied to the specimen through a Si cantilever and the crack propagates along the epoxy/glass interface. The epoxy film is the only creeping material in the specimen, so the crack growth is expected to be due to the creep of the epoxy.

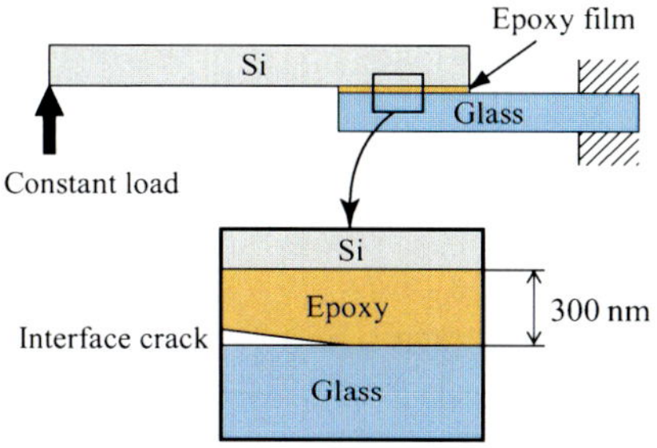

Figure 4.21 Specimen and testing system for a creep crack growth experiment. The target interface is between an epoxy thin film and a glass substrate. Crack growth can be observed through the glass substrate by microscopy.

Figure 4.22 shows the crack propagation curve under the constant load. The interface crack growth is time dependent, and this signifies the creep effect of the epoxy. The typical behavior observed is as follows: The crack growth rate is almost constant at an early stage, then it gradually accelerates as the crack propagates, and finally the crack unstably propagates, leading to complete delamination.

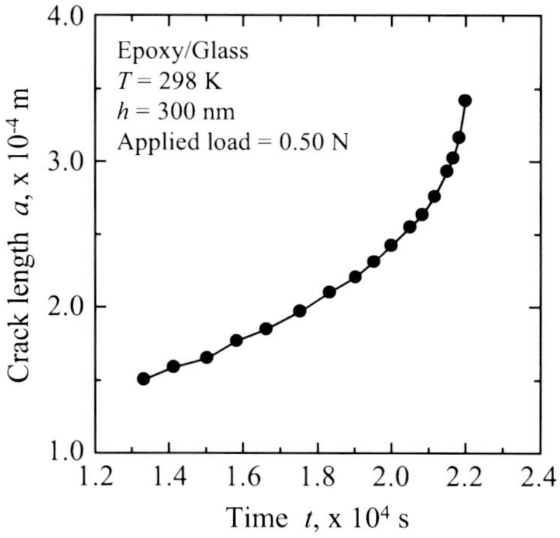

Figure 4.22 The relationship between crack length and time during a creep crack growth experiment. The crack grows along the interface due to the creep of the epoxy film.

Figure 4.23 shows the transition of normal stress distribution near the stationary interface crack in the specimen. Although the stress field transits from the elastic to creep singular fields, it takes more than 10^{15} s to reach the steady state (LSC), while the total loading time up to the complete delamination in the creep experiment is on the order of 10^4 s including the crack initiation time. Therefore, the crack grows before the stress field reaches the LSC condition. In such a case, it is likely that the crack grows under the SSC condition when the elastic stress field rather than the creep one is dominant.

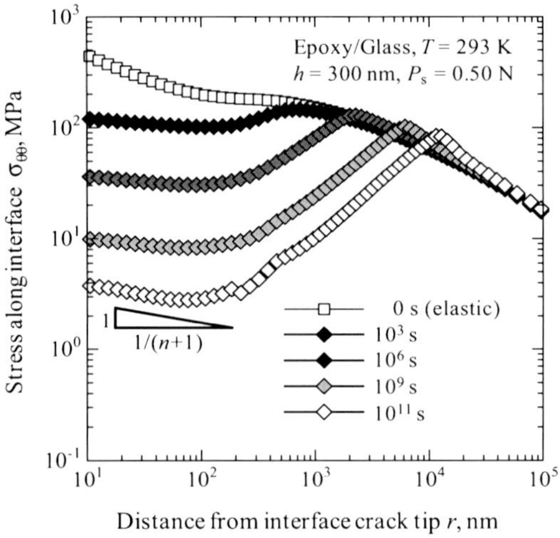

Figure 4.23 Transition of stress field near the interface crack tip between the epoxy thin film and the glass substrate. The intensity of the singular stress field decreases with the passage of time and the singularity transits from the elastic to the creep singularities.

Figure 4.24 shows the relationship between the crack growth rate, da/dt, and the elastic stress intensity factor, K_i. The result for a specimen with a 100 nm-thick epoxy film is also plotted on the figure. There is a good correlation between da/dt and K_i, indicating that the elastic singular stress field governs the creep crack growth; i.e., the crack growth under the SCC condition. The crack growth rate of the epoxy/glass interface is approximately expressed by the following relation.

$$\frac{da}{dt} = 33.2K_i^{9.2} \tag{4.7}$$

This behavior is similar to the creep crack growth under the SSC condition observed in a unidirectional carbon fiber reinforced plastic (CFRP), where the crack propagates along the fiber/matrix boundaries and the creep deformation is constrained by the fibers and confined to the vicinity of the crack tip [Ohtani et al., 1993; Uematsu et al., 1995].

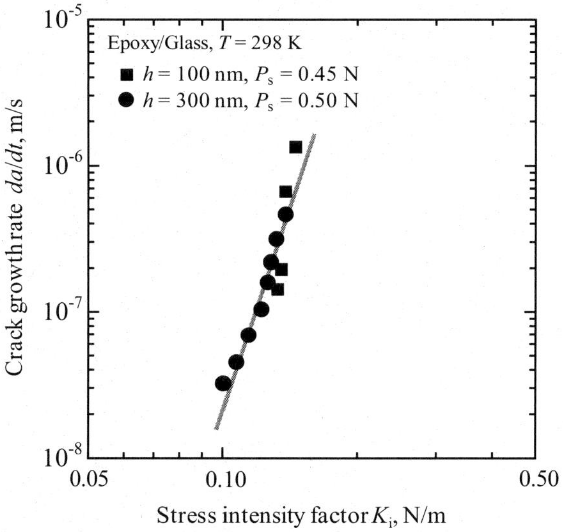

Figure 4.24 The relationship between crack growth rate and stress intensity factor. Good correlation exists between them, indicating that the elastic singular stress field governs the creep crack growth.

For nano-thick films, where creep deformation is more strongly restricted by the substrate, it takes longer for the stress field at the interface crack tip to reach the LSC condition than it does with submicron-thick films. The creep crack seems to grow under the SSC condition. On the other hand, because the size of the singular stress field proportionally scales down with the film thickness, the applicability of conventional mechanics governing the creep crack growth is questionable considering the creep mechanism at the atomic/molecular level.

4.2.3 Environment-Assisted Crack Growth

Environmental factors such as humidity greatly accelerate crack growth due to stress corrosion [Wiederhorn, 1967; Magnin 1995]. It has been reported that the interface crack between thin films is extremely affected by humidity [Dauskardt et al., 1998; Lane et al., 2001; Hughey et al., 2004; Xu et al., 1999]. A few studies on corrosion fatigue for several-micron-thick films have been conducted in moist air [Cannon et al., 1991; Kruzic et al., 2004; Gaudette et al., 1999] as pointed out in Section 4.2.1. It has been found that a crack grows along the interface between a 1500 nm-thick copper (Cu) film and a glass substrate in air with 45% relative humidity (R.H.) under cyclic loading, and the growth rate is many orders of magnitude faster than that under monotonic loading [Cannon et al., 1991].

(a)

(b)

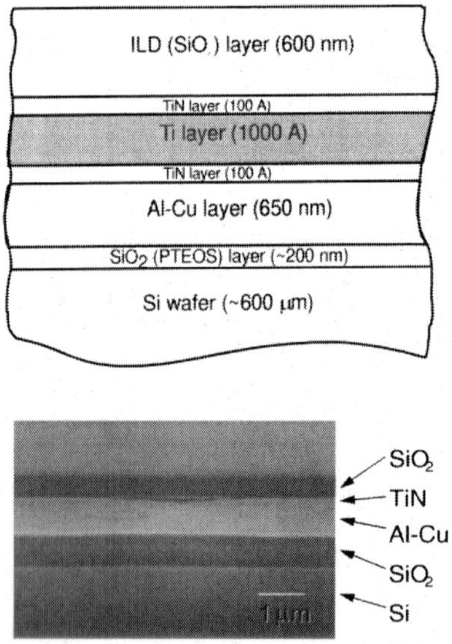

Figure 4.25 A multi-layer thin film interconnect structure showing (a) a schematic illustration with length scales and (b) a cross-section SEM micrograph. The target interface is between the TiN and SiO_2 films. Reprinted from Dauskardt, R. H., et al. (1998) "Adhesion and debonding of multi-layer thin film structures," *Eng. Fract. Mech.*, 61, pp. 141–162, Copyright (1998), with permission from Elsevier.

Figure 4.25 shows tested multi-layer thin films where metallic and dielectric films are stacked on a Si substrate and the target investigated is the interface between 600 nm-thick SiO_2 and 10 nm-thick TiN films [Dauskardt et al., 1998]. The experiments were conducted using the four-point bend method, where the sandwich structure was fabricated by diffusion bonding of Cu layers. The crack length was evaluated by a load-relaxation technique using constant displacements where the crack length was estimated by the linear-elastic relationship between the compliance of the four-point bend specimen and the crack length.

Figure 4.26a shows the measured crack length a as a function of time t at a humidity of ~45% R.H. and a temperature of 25°C. The crack grows time-dependently under the constant displacement condition and the crack growth rate gradually decreases with the passage of time due to the decrease in the crack driving force G. Since the specimen does not include creeping components such as polymers at room temperature, the crack growth along the interface is due to the moisture-assisted growth.

Figure 4.26b shows the relationship between the crack growth rate da/dt and the energy release rate G. The da/dt value has a good correlation to G, and the relation is expressed as the following equation.

$$\frac{da}{dt} = 2.2 \times 10^{-17} G^{14} \tag{4.8}$$

The environment-assisted crack grows below the fracture toughness of the interface G_c = 10.6 J/m^2.

Environment-assisted crack growth behavior in bulk materials typically consists of three stages: the threshold to intermediate region (Region I), the transport-controlled rate-limiting state (Region II), and the critical fracture (Region III). The above example seems to be in Region I, and the evaluation including Regions I and II for 25 nm-thick Ta and SiO_2 interfaces has also been reported, as shown in Fig. 4.27 [Lane et al., 2001].

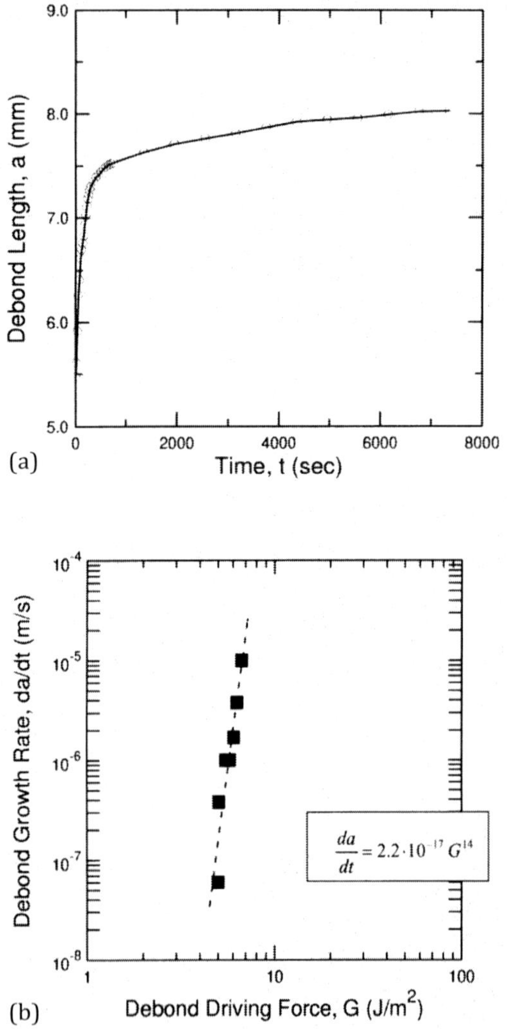

Figure 4.26 Typical load relaxation data showing (a) crack extension with the passage of time and (b) debond-growth rate da/dt as a function of the applied energy release rate G in laboratory air (R.H. ~ 45%) at 25°C. Reprinted from Dauskardt, R. H., et al. (1998) "Adhesion and debonding of multi-layer thin film structures," *Eng. Fract. Mech.*, 61, pp. 141–162, Copyright (1998), with permission from Elsevier.

The mechanisms of environmental-assisted crack growth in submicron or nano-thick components have not been clarified yet because the complex chemical reaction is accompanied at the crack tip. Although the mechanical issues for the applicability of fracture mechanics to the nano-scales are similar to those in fatigue and creep cracks, the dominant region size of the complex chemical reaction is unknown and few studies have been conducted.

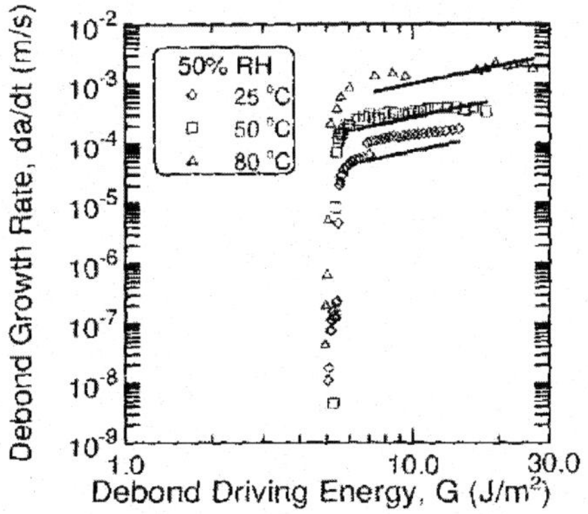

Figure 4.27 Subcritical debonding curves for failure at the interface between 25 nm-thick Ta and SiO_2 films for a range of environmental conditions. The crack growth behavior consists of Regions I and II, which is typically seen in environment-assisted crack growth in bulk materials. Reprinted from Lane, M. W., Snodgrass, J. M., and Dauskardt, R. H. (2001) "Environmental effects on interfacial adhesion," *Microelectron. Reliab.*, 41, pp. 1615–1624, Copyright (2001), with permission from Elsevier.

References

Al-Khanbashi, A., Petiau, C., Hamdy, A. E., and Moet, A. (2003) Fracture mechanics approach to forecast delamination lifetime, *J. Adhesion Sci. Technol.*, 17, pp. 1093–1107.

Bagchi, A., Lucas, G. E., Suo, Z., and Evans, A. G. (1994) A new procedure for measuring the decohesion energy for thin ductile films on substrates, *J. Mater. Res.*, 9, pp. 1734–1741.

Bagchi, A., and Evans, A. G. (1996) Measurement of the debond energy for thin film metallization lines on dielectrics, *Thin Solid Films*, 286, pp. 203–212.

Biner, S. B. (1997a) A numerical analysis of time-dependent stress fields ahead of stationary interface cracks at creep regime-Part I. Interface cracks in bi-material. *Eng. Fract. Mech.*, 56, pp. 513–529.

Biner, S. B. (1997b) A numerical analysis of time-dependent stress fields ahead of stationary interface cracks at creep regime-Part II. Interface cracks in layered materials. *Eng. Fract. Mech.*, 56, pp. 657–674.

Buchwalter, L. P. (2001) Relative adhesion measurement for thin film microelectronic structures. Part II, in *Adhesion Measurement of Films and Coatings* 2, ed. Mittal, K. L., VSP, Utrecht, pp. 19–47.

Cannon, R. M., Dalgleish, B. J., Dauskardt, R. H., Oh, T. S., and Ritchie, R. O. (1991) Cyclic fatigue-crack propagation along ceramic/metal interfaces, *Acta Metall. Mater.*, 39, pp. 2145–2156.

Charalambides, P. G., Lund, J., Evans, A. G., and Mcmeeking, R. M. (1989) A test specimen for determining the fracture resistance of bimaterial interfaces. *J. Appl. Mech.*, 111, pp. 77–82.

Dauskardt, R. H., Lane, M., Ma, Q., and Krishna, N. (1998) Adhesion and debonding of multi-layer thin film structures, *Eng. Fract. Mech.*, 61, pp. 141–162.

Dauskardt, R. H., and Lane, M. (2000) Adhesion and reliability of opper. interconnects with Ta and TaN barrier layers, *J. Mater. Res.*, 15, pp. 203–211.

De Boer, M. P., Kriese, M. D., and Gerberich, W. W. (1997) Investigation of a new fracture mechanics specimen for thin film adhesion measurement, *J. Mater. Res.*, 12, pp. 2673–2685.

Erdogan, F. (1963) Stress distribution in a nonhomogeneous elastic plane with cracks, *J. Appl. Mech.*, 30, pp. 232–238.

Evans, A. G., and Dalgleish, B. J. (1992) The fracture resistance of metal-ceramic interfaces, *Acta Metall. Mater.*, 40, pp. S295–306.

Evans, A. G., Hutchinson, J. W., and Wei, Y. (1999) Interface adhesion: effect of plasticity and segregation, *Acta Mater.*, 47, pp. 4093–4113.

Fleck, N. A., Kang, K. J., and Ashby, M. F. (1994) The cyclic properties of engineering materials, *Acta Metall. Mater.*, 42, pp. 365–381.

Gao, L., Nakasa, K., Kato, M., and Nishida, H. (2003) Evaluation of interfacial fracture toughness of thermal barrier coating under heat cycles. *Key Eng. Mater.*, 243, pp. 267–272.

Gaudette, F., Suresh, S., and Evans, A. G. (1999) The influence of solid-state and liquid-phase bonding on fatigue at Al/Al2O3 interfaces, *Metall. Mater. Trans.*, 30A, pp. 763–769.

Gerberich, W. W., Kramer, D. E., Tymiak, N. I., Volinsky, A. A., Bahr, D. F., and Kriese, M. D. (1999), Nanoindentation-induced defect-interface interactions: phenomena, methods and limitations, *Acta Mater.* 47, pp. 4115–4123.

Hay, J. C., Liniger, E. G., and Liu, X. H. (2001) Measurement of interfacial fracture energy in microelectronic multifilm applications, in *Adhesion Measurement of Films and Coatings* 2, ed. Mittal, K. L., VSP, Utrecht, pp. 205–217.

He, M. Y., Evans, A. G., and Hutchinson, J. W. (1997) Convergent debonding of films and fibers, *Acta Mater.*, 45, pp. 3481–3489.

Hirakata H., Kitamura, T., and Kusano, T. (2005) Pre-cracking technique for fracture mechanics experiment along interface between thin film and substrate, *Eng. Fract. Mech.*, 72, pp. 1892–1904.

Hirakata, H., Kitazawa, M. and Kitamura, T. (2006) Fatigue crack growth along interfacial between metal and ceramics submicron-thick films in inert environment, *Acta Mater.*, 54, pp. 89–97.

Hirakata, H., Takahashi, Y., Truong, D. V., and Kitamura, T. (2007) Role of plasticity on interface crack initiation from a free edge and propagation in a nano-component, *Int. J. Fract.*, 145, pp. 261–271.

Hohfelder, R., Luo, H., Vlassak, J., Chidsey, C., and Nix, W. (1997) Measuring interfacial fracture toughness with the blister test, *MRS Symp. Proc.*, **436, pp.** 115–120.

Hughey, M. P., Morris, D. J., Cook, R. F., Bozeman, S. P., Kelly, B. L., Chakravarty, S. L. N., Harkens, D. P., and Stearns, L. C. (2004) Four-point bend adhesion measurements of copper and permalloy systems, *Eng. Fract. Mech.*, 71, pp. 245–261.

Hutchinson, J. W., and Suo, Z. (1992) Mixed mode cracking in layered materials, *Adv. Appl. Mech.*, 29, pp. 63–191.

Ikeda, T., Miyazaki, N., and Soda, T. (1998) Mixed mode fracture criterion of interface crack between dissimilar materials, *Eng. Fract. Mech.*, 59, pp. 725–735.

Jensen, H. M. (1991) The blister test for interface toughness measurement, *Eng. Fract. Mech.*, 40, pp. 475–486.

Kamiya, S., Kimura, H., Yamanobe, K., Saka, M., and Abe, H. (2002) A new systematic method of characterization for the strength of thin films on substrates-evaluation of mechanical properties by means of 'film projection,' *Thin Solid Films*, 414, pp. 91–98.

Kriese, M. D., Gerberich, W. W., and Moody, N. R. (1999a) Quantitative adhesion measures of multilayer films: Part I. Indentation mechanics, *J. Mater. Res.*, 14, pp. 3007–3018.

Kriese, M. D., Gerberich, W. W., and Moody, N. R. (1999b) Quantitative adhesion measures of multilayer films: Part II. Indentation of W/Cu, W/W, Cr/W, *J. Mater. Res.*, 14, pp. 3019–3026.

Kruzic, J. J., McNaney, J. M., Cannon, R. M., and Ritchie, R. O. (2004) Effects of plastic constraint on the cyclic and static fatigue behavior of metal/ceramic layered structures, *Mech. Mater.*, 36, pp. 57–72.

Landes J. D., and Begley J. A. (1976) A fracture mechanics approach to creep crack growth, *ASTM STP 590*, pp.128–148.

Lane, M. W., Snodgrass, J. M., and Dauskardt, R. H. (2001) Environmental effects on interface adhesion, *Microelectronics Reliability*, 41, pp. 1615–1624.

Lane, M. (2003) Interface fracture, *Annu. Rev. Mater. Res.* 33, pp. 29–54.

Magnin, T. (1995) Recent advances for corrosion fatigue mechanisms, *ISIJ Int.* 35, pp. 223–233.

Marshall, D. B., and Evans, A. G. (1984) Measurement of adherence of residually stressed thin films by indentation. I. Mechanics of interface delamination, *J. Appl. Phys.*, 56, pp. 2632–2638.

McNaney, J. M., Cannon, R. M., and Ritchie, R. O. (1996) Fracture and fatigue-crack growth along aluminum-alumina interfaces, *Acta Mater.*, 44, pp. 4713–4728.

Minoshima, K., and Komai, K. (1988) Stress corrosion cracking of high-strength steels under dynamic loading conditions, in *Localized Corrosion, Current Japanese Materials Research 4*, ed. Hine, F., Komai, K., and Yamakawa, K., Elsevier, Amsterdam, pp. 123–147.

Mutoh, Y., Xu, J. Q., Miyashita, Y., Bernardo, G. G., and Takahashi, M. (2003) Interface fatigue crack propagation in ceramic thermal barrier coating, *JSME Int. J. Ser. A*, 46, pp. 403–408.

Nied, H. F. (2003) Mechanics of interface fracture with applications in electronic packaging, *IEEE Transactions on Device and Materials Reliability*, 3, pp. 129–143.

Ohtani, R., Kitamura, T., Uematsu, Y., and Hojo, M. (1993) ASTM STP 1174, ed. Harris, C. E., and Gates, T. S., Mode I delamination of carbon fiber reinforced thermoplastic polymer under static and cyclic creep at elevated temperatures, in *High Temperature and Environmental Effects on*

Polymeric Composites, American Society for Testing and Materials, Philadelphia, pp. 23–38.

Okazaki, M., Okamoto, M., and Harada, Y. (2001) Interfacial fatigue crack propagation in Ni-base superalloy protective coatings, *Fatigue Fract. Eng. Mater. Struct.*, 24, pp. 855–865.

Omiya, M., Inoue, H., and Kishimoto, K. (2004) Evaluation of interfacial strength by multi-stages peel test, *Key Eng. Mater.*, 261–263, pp. 483–488.

Reimanis, I. E., Dalgleish, B. J., Brahy, M., Ruhle, M., and Evans, A. G. (1990) Effects of plasticity on the crack propagation resistance of a metal/ceramic interface, *Acta Metall. Mater.*, 38, pp. 2645–2652.

Rice, J. R. (1968) A path independent integral and the approximate analysis of strain concentration by notches and cracks, *J. Appl. Mech.*, 35, pp. 379–386.

Rice, J. R. (1988) Elastic fracture mechanics concepts for interface cracks, *J. Appl. Mech.*, 55, pp. 98–103.

Rolfe, S. T., and Barsom, J. M. (1977) *Fracture and fatigue control in structures*, Prentice-Hall, Upper Saddle River, NJ.

Rossington, C., Evans, A. G., Marshall, D. B., and Khuri-Yakub, B. T. (1984) Measurement of adherence of residually stressed thin films by indentation. II. Experiments with ZnO/Si, *J. Appl. Phys.*, 56, pp. 2639–2644.

Schwaiger, R., Dehm, G., and Kraft, O. (2003) Cyclic deformation of polycrystalline Cu films, *Philos. Mag.*, 83, pp. 693–710.

Schwaiger, R., and Kraft, O. (2003) Size effects in the fatigue behavior of thin Ag films, *Acta Mater.*, 51, pp. 195–206.

Sekiguchi, A., and Koike, J. (2008) Evaluation of interface adhesion strength in Cu/(Ta-x% N, Ta/TaN)/SiO2/Si by nanoscratch test, *Japn. J. Appl. Phys.*, 47, pp. 1042–1049.

Shaw, M. C., Marshall, D. B., Dalgleish, B. J., Dadkhah, M. S., He, M. Y., and Evans, A. G. (1994) Fatigue crack growth and stress redistribution at interfaces, *Acta Metall. Mater.*, 42, pp. 4091–4099.

Taira, S., Ohtani, R., and Kitamura, T. (1979) Application of J-integral to high-temperature crack propagation, Part I—Creep crack propagation, *Trans. ASME Ser. H*, 101, pp. 154–161.

Takahashi, Y., Hirakata, H., and Kitamura, T. (2007) Quantitative evaluation of plasticity of a ductile nano-component, *Thin Solid Films* 516, pp. 1925–1930.

Thouless, M. D. (1994) Fracture mechanics for thin-film adhesion, *IBM J. Res. Develop.,* 38, pp. 367–377.

Truong, D. V., Hirakata, H., and Kitamura, T. (2006) Effect of loading frequency on fatigue crack growth between a submicron-thick film and a substrate, *JSME Int. J. Ser. A*, 49, pp. 370–375.

Tymiak, N. I., Volinsky, A. A., Kriese, M. D., Downs, S. S., and Gerberich, W. W. (1999) The role of plasticity in bimaterial fracture with ductile interlayers, *Metall. Mater. Trans. A*, 31, pp. 863–872.

Uematsu, Y., Kitamura, T., and Ohtani, R. (1995) Delamination behavior of a carbon-fiber-reinforced thermoplastic polymer at high temperatures, *Composites Science and Technology,* 53, pp. 333–341.

Venkataraman, S., Kohlstedt, D. L., and Gerberich, W. W. (1996) Continuous microscratch measurements of the practical and true works of adhesion for metal/ceramic system, *J. Mater. Res.*, 11, pp. 3133–3145.

Volinsky, A. A., Moody, N. R., and Gerberich, W. W. (2002) Interface toughness measurements for thin films on substrates, *Acta Mater.*, 50, pp. 441–466.

Wiederhorn, S. M. (1967) Influence of water vapor on crack propagation in soda-lime glass, *J. Am. Ceram. Soc.*, 50, pp. 407–414.

Wang, J. S., and Suo, Z. (1990) Experimental determination of interfacial toughness curves using Brazil-Nuts-Sandwiches, *Acta Metal. Mater.*, 38, pp. 1279–1290.

Xu, G., He, M. Y., and Clarke, D. R. (1999) The effect of moisture on the fracture energy of TiN/SiO2 interface in multi-layer thin films, *Acta Mater.*, 47, pp. 4131–4141.

Zhang, G. P., Volkert, C. A., Schwaiger, R., Wellner, P., Arzt, E., and Kraft, O. (2006) Length-scale-controlled fatigue mechanisms in thin copper films, *Acta Mater.*, 54, pp. 3127–3139.

Zhuk, A. V., Evans, A. G., Hutchinson, J. W., and Whitesides, G. M. (1998) The adhesion energy between polymer thin films and self-assembled monolayers, *J. Mater. Res.*, 13, pp. 3555–3564.

Chapter 5

INITIATION OF INTERFACE CRACK

5.1 THE SMALLEST LIMIT OF FRACTURE MECHANICS CONCEPT

An interface edge is the likeliest site of fracture initiation because of the stress singularity due to modulus mismatch. As illustrated in Fig. 5.1, the singular stress (stress-concentrated) region near the interface edge proportionally scales down [Becker et al., 1997] in a continuum media as the size of a component becomes small. In a component with single-micrometer or nanometer size, the region governed by the singular field is confined in the nano-scale [Kitamura et al., 2002, 2003]. In such a case, the applicability of fracture mechanics to the crack initiation from the edge is questionable because the concept is built on continuum mechanics.

In this chapter, we explore the smallest limit of conventional fracture mechanics for a stress singular (concentration) region on the basis of a series of experimental studies on crack initiation at interface edges in submicron and nanometer size components consisting of films or islands on a substrate.

Although a number of studies of crack propagation have been conducted, few efforts have been devoted to crack initiation at interface edges in micron or smaller components. Crack initiation is a more important phenomenon than crack propagation because the initiation of a tiny crack immediately leads to a fatal malfunction

Fracture Nanomechanics
Edited by Takayuki Kitamura, Hiroyuki Hirakata, Takashi Sumigawa and Takahiro Shimada

www.panstanford.com

of the device. Some new techniques are necessary for this inquiry because no conventional experimental methods are available.

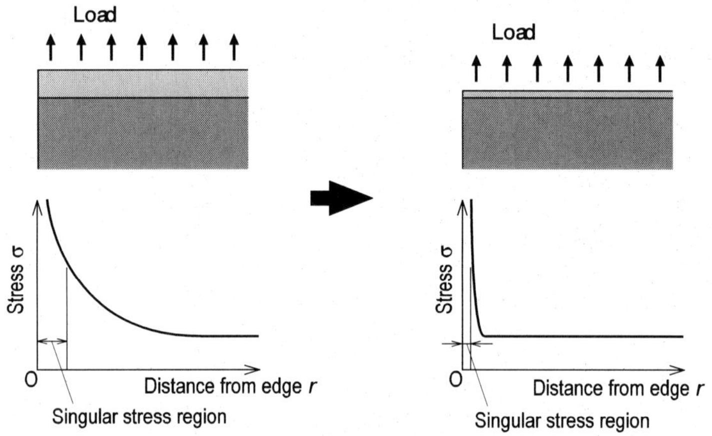

Figure 5.1 Size dependence of the singular stress region at an interface edge between a thin film and a substrate.

5.2 TOUGHNESS

5.2.1 Crack Initiation at the Interface Edge between Thin Film and Substrate

Figure 5.2 shows two sandwich specimens, designated Type A and Type B, which have similar corner angles of 90°/90° at the interface edge [Kitamura et al., 2002, 2003]. Thus, the order of stress singularity λ near the edge is identical in the specimens, as explained in Section 2.5.1. However, the stress distribution away from the interface edge is significantly different between them because of the different types of loading (push-up in Type A; push-down in Type B). If the crack initiation is governed by the singular stress field, the critical stress intensity must be the same in both the types. Here, since the shear stress is much smaller than the tensile stress in these specimens, the interface crack initiates under the Mode I condition in both types.

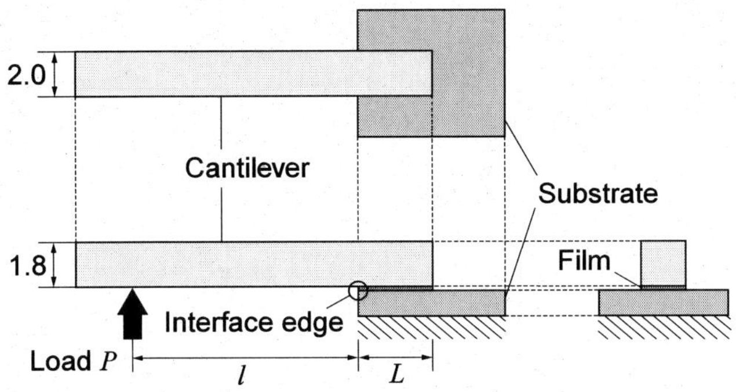

(a) Type A.

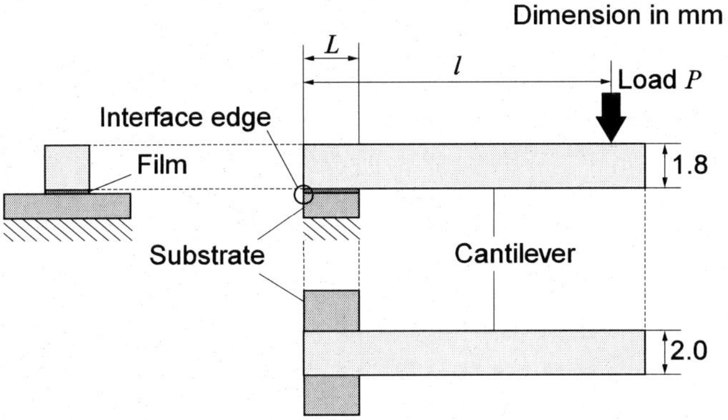

(b) Type B.

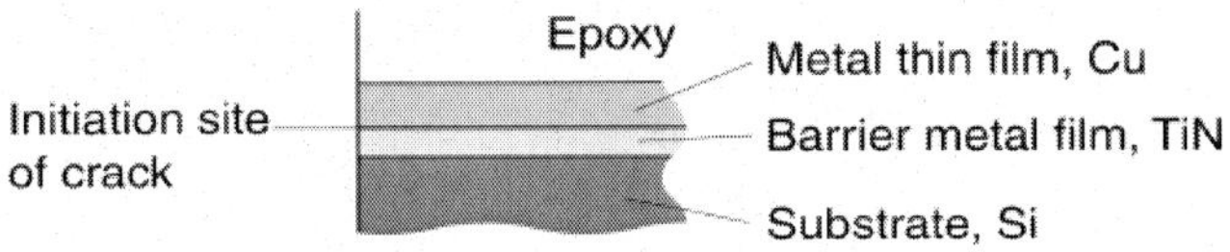

(c) Detail of interface edge; an example for Cu/TiN thin films.

Figure 5.2 Cantilever specimens for crack initiation experiments. Reprinted from Kitamura, T., Hirakata H., and Itsuji T. (2003) "Effect of residual stress on delamination from interface edge between nano-films," *Eng. Fract. Mech.*, 70, pp. 2089–2101, Copyright (2003), with permission from Elsevier.

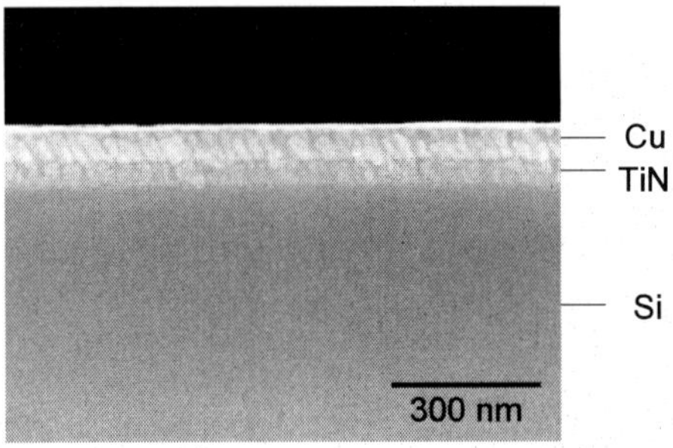

Figure 5.3 FE-SEM micrograph of a cross-sectioned surface near the interface between Cu (thickness of 50 nm) and TiN (thickness of 50 nm) films Reprinted from Kitamura, T., Hirakata H., and Itsuji T. (2003) "Effect of residual stress on delamination from interface edge between nano-films," *Eng. Fract. Mech.*, 70, pp. 2089–2101, Copyright (2003), with permission from Elsevier.

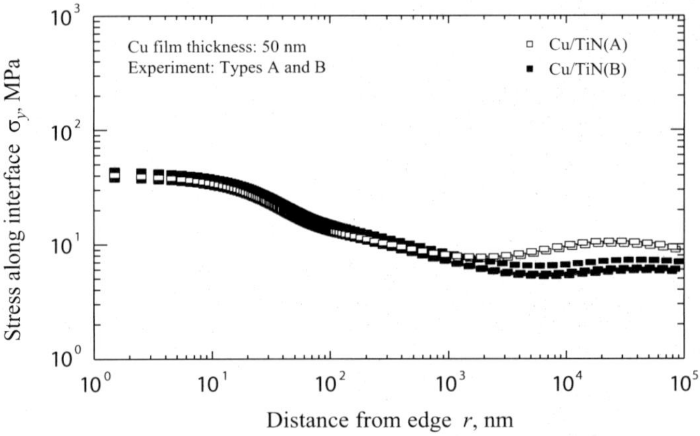

Figure 5.4 Stress distributions near the interface edge between Cu/TiN thin films at crack initiation Reprinted from Kitamura, T., Hirakata H., and Itsuji T. (2003) "Effect of residual stress on delamination from interface edge between nano-films," *Eng. Fract. Mech.*, 70, pp. 2089–2101, Copyright (2003), with permission from Elsevier.

Figure 5.3 shows the edge of tested material consisting of 50 nm-thick films (Cu and TiN) on a Si substrate. The load-displacement curve is linear up to the delamination, where the load sharply drops. Observation on the fracture surface by means of an electron microscope and Auger electron spectroscopy (AES) reveals that the crack is initiated and propagates perfectly along the Cu/TiN interface. Figure 5.4 shows elastic normal stress distributions near the Cu/TiN interface edge at the crack initiation, which is analyzed under the plane strain condition by the boundary element method (BEM). The stress field has the form

$$\sigma_{ij} = \frac{K_{\mathrm{f}}}{r^{\lambda}} f_{ij}(\theta) \tag{5.1}$$

near the interface edge, and the stress singularity λ is 0.02 where the region size of the singular field is on the order of 10 nm. The result signifies that the stress in Types A and B coincides well in the region until about 500 nm, including the singular stress field, while the distributions are much different outside the region, $r > 500$ nm. It suggests that the singular stress field governs crack initiation. However, we cannot exclude a hypothesis that the stress in about 500 nm dominates the crack initiation.

Other evidence for various interface edges of submicron-thick films having different stress singularity [Kitamura et al., 2002, 2007a] and piezoelectric films [Shang et al., 2005a, 2005b] supports the applicability of the fracture mechanics concept using K_{f}. Some examples are listed in Table 5.1.

Table 5.1 Examples of toughness evaluation of interface edges of thin films

Materials	**Thickness, nm**	**Singularity λ**	**Toughness K_{fC}, MPa m$^{-\lambda}$**	**Reference**
Cu/TiN	50/50	0.02	25	Kitamura et al., 2003
Cu/TiN	500/50	0.02	11	Kitamura et al., 2003
Cu/TaN	300/50	0.08	9.0	Kitamura et al., 2002
Cr/PZT	200/2500	0.334	2.21	Shang et al., 2005a, 2005b
Al/Si	200/substrate	0.069	54.7	Kitamura et al., 2007a
Al/SiO_2	200/100	0.011	159.7	Kitamura et al., 2007a
Au/Si	200/substrate	0.067	5.3	Kitamura et al., 2007a
Cu/Si	200/sybstrate	0.009	64.0	Kitamura et al., 2007a
Cu/Ta	200/50	0.014	77.0	Kitamura et al., 2007a

Figure 5.5 shows the stress distributions at the crack initiation of the two Cu/TiN specimens where the thicknesses of Cu are 50 nm and 500 nm, respectively. Attention, however, must be paid to the fact that the stress near the interface edge is strongly dependent on the film thickness. Since the deposition process on the Cu film was the same except for the thickness, the interface must have the same structure and the same toughness. The difference between the magnitudes of K_f suggests that other stress sources have acted on the interface edge.

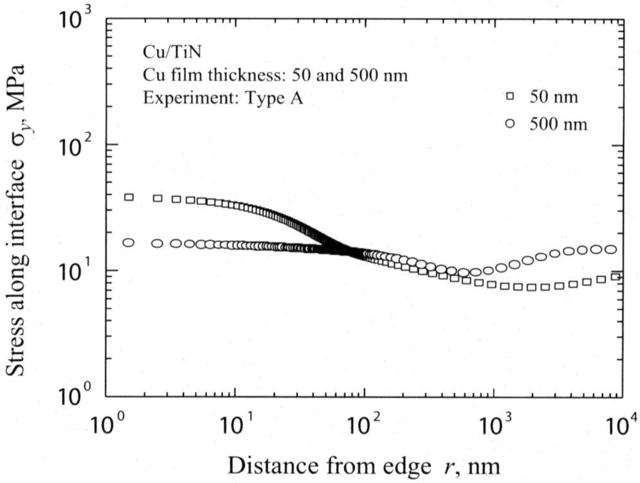

Figure 5.5 Effect of film thickness on the critical stress distribution near the interface edge between Cu and TiN films Reprinted from Kitamura, T., Hirakata H., and Itsuji T. (2003) "Effect of residual stress on delamination from interface edge between nano-films," *Eng. Fract. Mech.*, 70, pp. 2089–2101, Copyright (2003), with permission from Elsevier.

It is well known that residual stress is introduced in a thin film during fabrication. Figure 5.6 shows the stress distributions near the Cu/TiN interface edges induced by the residual stress of 50 and 500 nm-thick Cu films which are σ_r = 186 MPa and 250 MPa on average, respectively [Kitamura et al., 2003]. The residual stress is remarkably intensified near the interface edge due to the modulus mismatch even without the external loading. The thicker film has a higher intensified stress than the thinner film.

Figure 5.7 shows the critical stress distributions near the Cu/TiN interface edge at the delamination superimposing the contributions of the residual stress (Fig. 5.6) and the applied load (Fig. 5.5). The stress singularity λ analyzed by the BEM is 0.18, which is higher than that under the external load without the residual stress; $\lambda = 0.02$ (see Figs. 5.4 and 5.5). The stress intensity near the interface edge shows a good agreement regardless of film thickness. This signifies that, considering not only the external load but also the residual stress, we can evaluate the crack initiation strength as a general criterion independently of loading type and film thickness. Since the residual stress works as a part of the driving force for crack initiation, the true driving force is the summation of stress intensities due to both external stress and residual stress.

The dominant region of singular stress fields is about 10 nm for the 50 nm-thick films and several tens of nm for the 500 nm-thick films as shown in Fig. 5.7. Although the region size of the singular stress field is different depending on the film thickness, both the stress intensities agree well in the region of about 20 nm. This indicates that the intensified (or concentrated) stress field of about 20 nm governs the crack initiation.

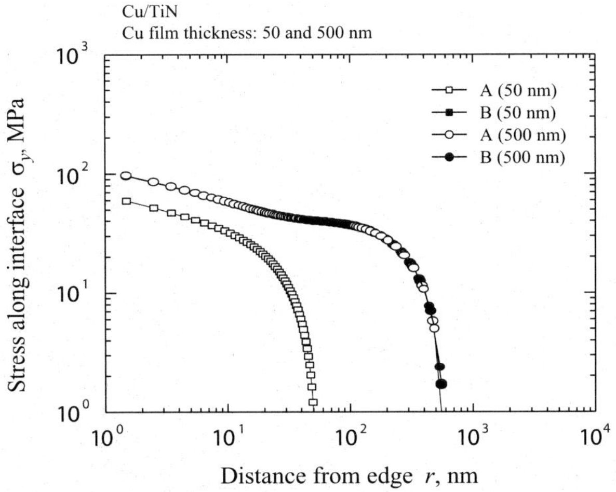

Figure 5.6 Stress distributions near the interface edge between Cu and TiN films at delamination due to the contributions of residual stress and external load.

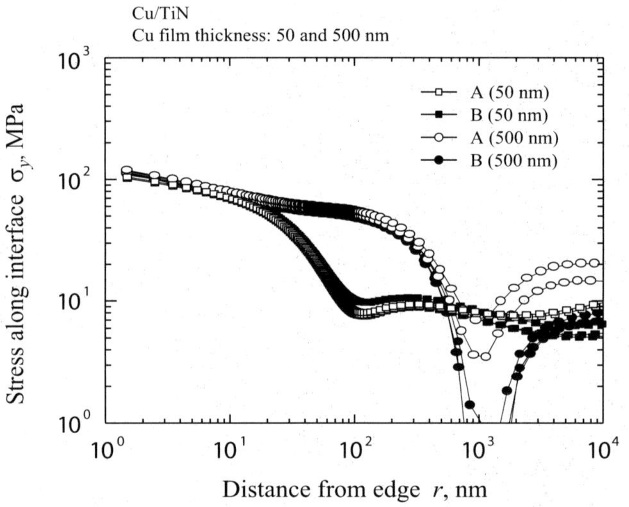

Figure 5.7 Stress distributions near the interface edge between Cu and TiN films at delamination due to the contributions of residual stress and external load. Reprinted from Kitamura, T., Hirakata H., and Itsuji T. (2003) "Effect of residual stress on delamination from interface edge between nano-films," *Eng. Fract. Mech.*, 70, pp. 2089–2101, Copyright (2003), with permission from Elsevier.

5.2.2 Crack Initiation at Interface Edge of an Island on a Substrate

In the cantilever method described in the previous section, a conventional experimental apparatus for material testing can be used, although cantilevers and adhesives are required to construct the specimen. A limitation of the method is that it cannot be applied to strong interfaces due to the presence of the weak adhesive layers and their interfaces. Moreover, the specimen size is on the millimeter range, while the target singular stress field is on the order of 10 nm ~ 100 nm. Because thin films have a macroscopic dimension, except the film thickness, the stress field and strength are averaged values over the long interface edge on the width of a two-dimensional specimen. This may cause experimental errors. So, it is important to consider whether fracture mechanics can be applied to the crack initiation of three-dimensional components in the nano-size. Most microelectronic and micromechanical devices consist of nano-components extended three-dimensionally, such as dots or islands. For nano-islands, there have been no experimental techniques proposed in terms of the interface strength.

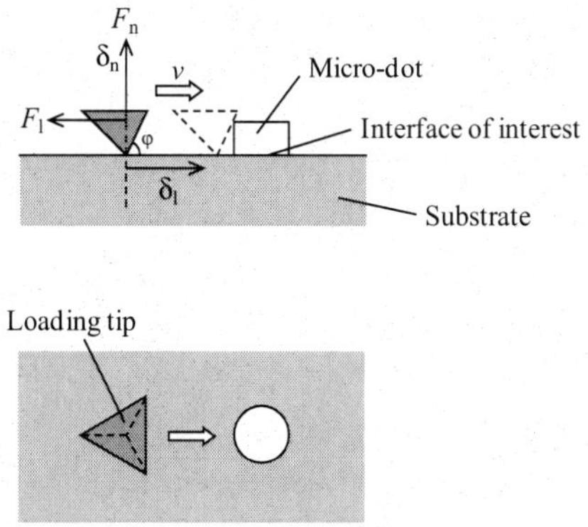

(a) Schematic illustration of loading experiment for an island on substrate.

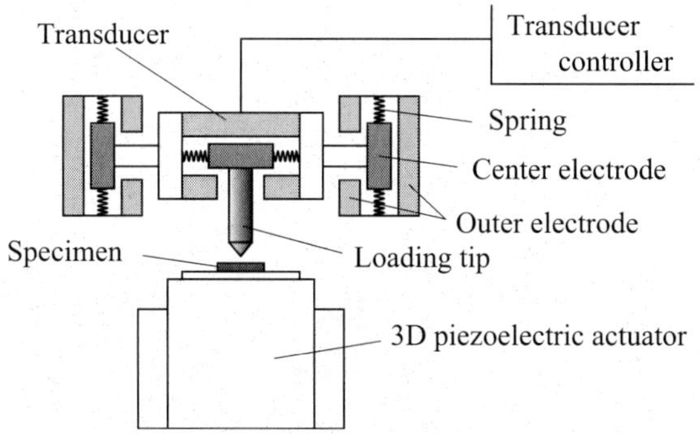

(b) Loading apparatus (Hysitron Inc., TriboScope).

Figure 5.8 Schematic illustration of a loading experiment for an island on substrate and loading apparatus. Part (a) reprinted from Hirakata, H., et al. (2006) "Dominant stress region for crack initiation at interface edge of microdot on a substrate," *Eng. Fract. Mech.*, 73, pp. 2698–2709, Copyright (2006), with permission from Elsevier. Part (b) reprinted from Hirakata, H., Kitamura, T., and Yamamoto, Y. (2004) "Direct measurement of interface strength between copper submicron-dot and silicon dioxide substrate," *JSME Int. J. Ser. A*, 47, pp. 324–330.

Figure 5.8a schematically illustrates a delaminating method for an island from a substrate [Hirakata et al., 2004a, 2004b, 2006]. A load is directly applied to the island by a diamond tip using a lateral loading apparatus built onto an atomic force microscope (AFM). The diamond tip, which has a tetrahedral shape, is dragged horizontally on the substrate surface, and the critical lateral load, F_{1C}, at the delamination is measured. Figure 5.8b shows the testing apparatus that can independently measure/control both the lateral and the normal forces and the displacements of the loading tip during the test. It consists of three force/displacement transducers, each of which has two fixed outer plates and a spring-suspended center plate. The loading tip is attached to the center plate of the middle transducer. The normal force of the tip is electrostatically generated by applying a voltage between the center and the lower plates in the middle transducer, and the resulting normal displacement is precisely detected by a change in the capacitance of the outer electrodes. The side transducers generate a lateral force in the same manner and control the lateral displacement using the capacitance signal.

Figure 5.9a shows the relationship between the lateral load and the displacement obtained by the experiment of a Cr island on the SiO_2/ Si substrate shown in Fig. 5.9b. After the loading tip hits the island, the load monotonically increases until it drops sharply at the point "a" due to the delamination. In addition, the direct loading to the island enables the delamination along a strong interface, whereas the use of sandwich methods is limited to relatively weaker interfaces due to the strength of the necessary adhesive layer.

This method cannot be applied to soft materials such as ductile metals and polymers because the contact zone collapses before delamination, as shown in Fig. 5.10. Figure 5.11 shows a modified method for a ductile island [Hirakata et al., 2004b]. A hard layer is deposited on the islands as a superlayer and the load is applied to it. The superlayer restrains the deformation and prevents the collapse of the island. Moreover, the influence of a complicated stress field resulting from the contact of the tip is eliminated because the contact zone is away from the interface edge where the delamination crack is initiated. Figure 5.12 shows AFM images after delamination experiments for a ductile Cu island with a W superlayer and the delaminated surface of the substrate. The absence of damage and plasticity on the W layer of the separated island points out that it works well as the constraint and stress conductive layer of Cu.

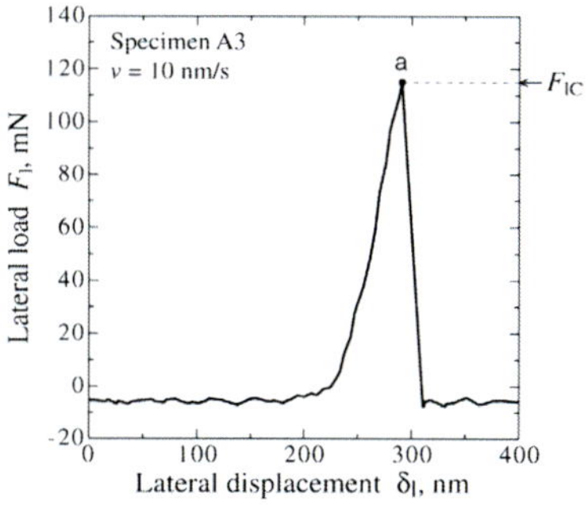

(a) Relationship between the lateral load and displacement during the experiment.

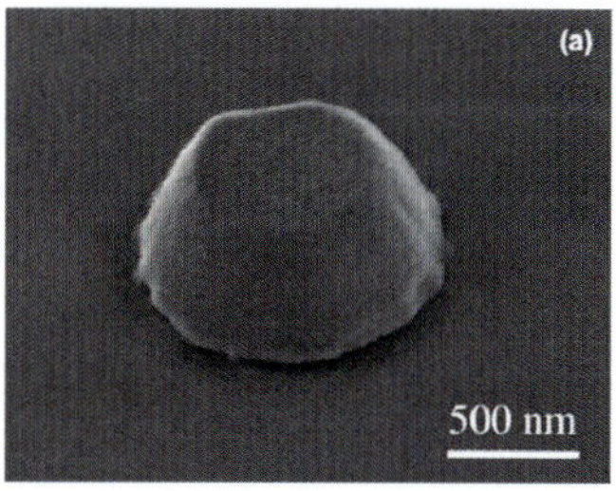

(b) Cr submicron island on SiO_2 substrate.

Figure 5.9 Delamination experiment for a Cr submicron island on SiO_2 substrate . Reprinted from Hirakata, H., et al. (2006) "Dominant stress region for crack initiation at interface edge of microdot on a substrate," *Eng. Fract. Mech.*, 73, pp. 2698–2709, Copyright (2006), with permission from Elsevier.

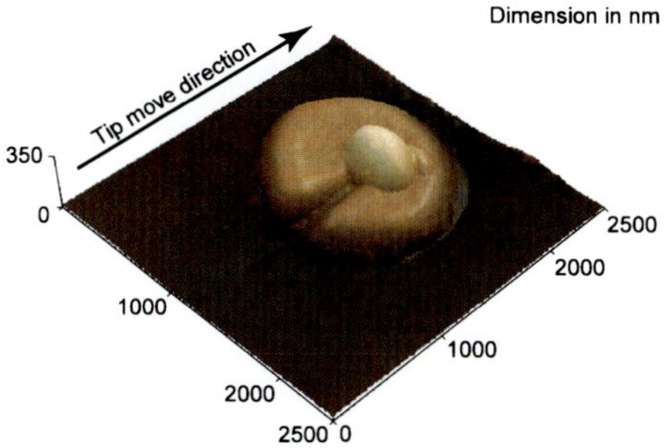

Figure 5.10 AFM image of a collapsed Cu island due to direct loading by the tip. Reprinted from Hirakata, H., Kitamura, T., and Yamamoto, Y. (2004) "Direct measurement of interface strength between copper submicron-dot and silicon dioxide substrate," *JSME Int. J. Ser. A*, 47, pp. 324–330.

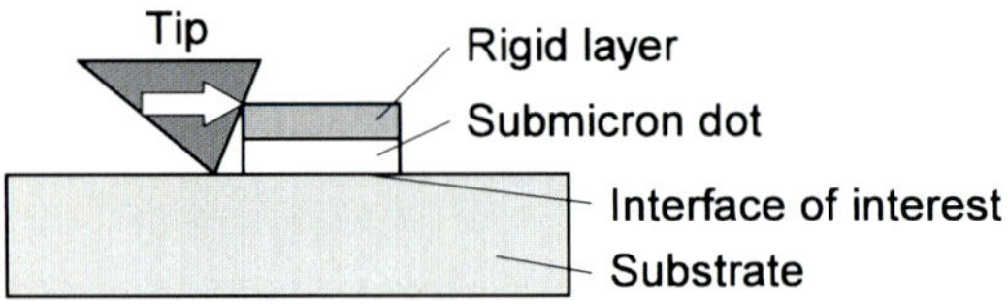

Figure 5.11 Method for delaminating a ductile island from a substrate.

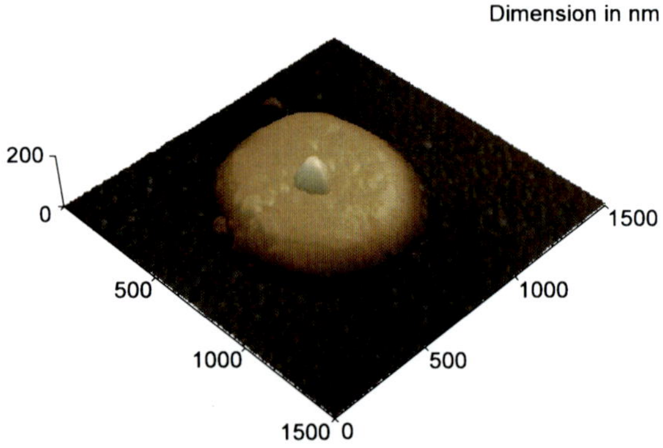

(a) Separated W/Cu island.

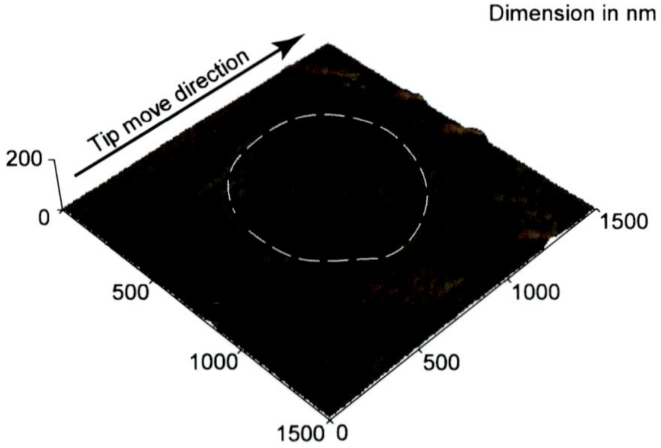

(b) Substrate surface.

Figure 5.12 AFM images of a separated ductile island with the superlayer and fracture surface. Reprinted from Hirakata, H., Kitamura, T., and Yamamoto, Y. (2004) "Direct measurement of interface strength between copper submicron-dot and silicon dioxide substrate," *JSME Int. J. Ser. A*, 47, pp. 324–330.

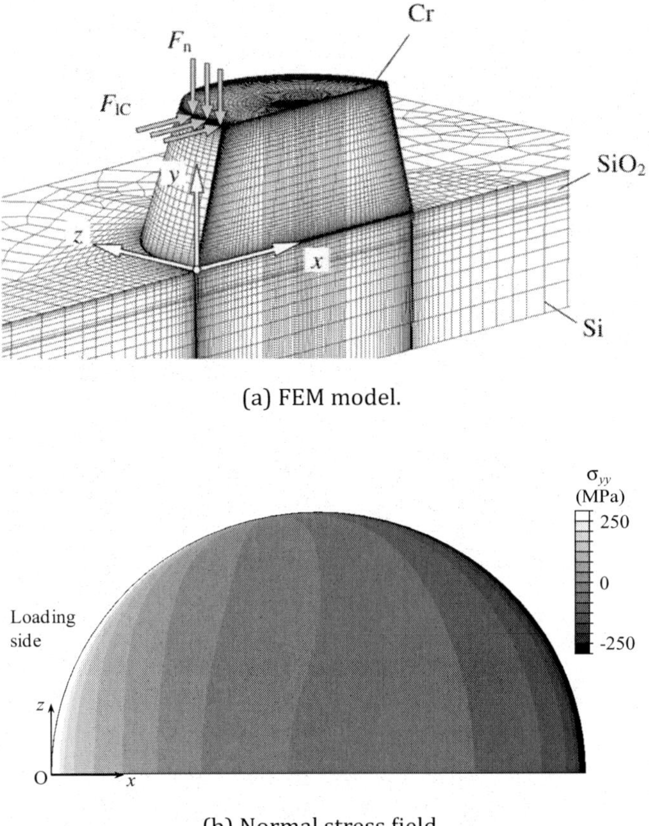

Figure 5.13 FEM model and stress distribution on the interface between a Cr island and SiO_2 substrate at delamination. Reprinted from Hirakata, H., et al. (2006) "Dominant stress region for crack initiation at interface edge of microdot on a substrate," *Eng. Fract. Mech.*, 73, pp. 2698–2709, Copyright (2006), with permission from Elsevier.

Figure 5.13b shows the normal stress distribution on the interface between the Cr island and SiO_2 substrate at the delamination (at "a" in Fig. 5.9a) calculated by a three-dimensional finite element method using the model of Fig. 5.13a. The tensile stress is highly concentrated on the loading side (left side in the figure), and the singular stress field exists near the rounded interface edge.

Here, consider the propagation of a two-dimensional crack with a round front in a three-dimensional body, e.g., the half elliptical crack in a bulk shown in Fig. 5.14a. Conventional fracture mechanics tells us that the crack growth is governed by the singular stress field on the normal plane to the crack front. If the concept is extended to the round interface edge of the island, in a manner similar to that shown in Fig. 5.14b, the crack initiation can be evaluated by the intensity of the singular stress field on the plane normal to the round interface edge.

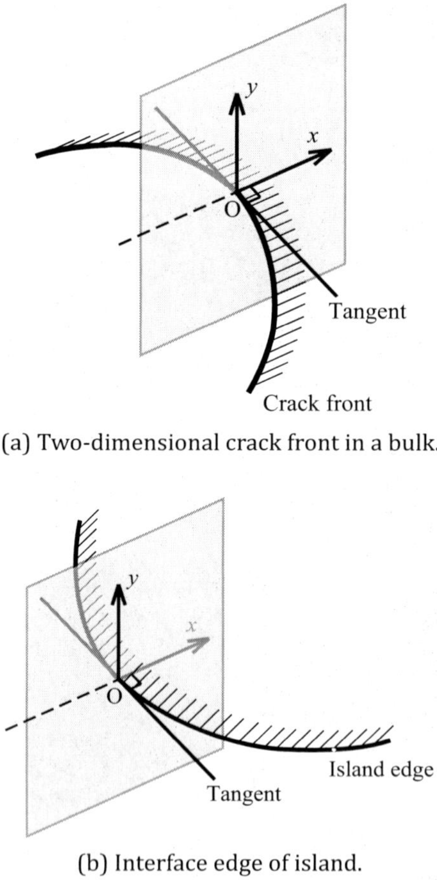

(a) Two-dimensional crack front in a bulk.

(b) Interface edge of island.

Figure 5.14 Schematic illustration showing definitions of a local coordinates system for a round crack tip and the round interface edge of an island.

Figure 5.15 shows the magnified distributions of normal and shear stresses, σ_{yy} and τ_{xy} near the island/substrate interface edge along the x axis. The stress distributions are almost straight in the double logarithmic chart, as shown the figure. Assuming the stress fields to be expressed by the following

$$\sigma_{yy}(r)=\frac{K_{\sigma}}{r^{\lambda_{\sigma}}},\tau_{xy}(r)=\frac{K_{\tau}}{r^{\lambda_{\tau}}} \quad (5.2)$$

the stress intensity is nominally described by K_σ and K_τ, where λ_σ and λ_τ are the apparent orders of singularity or the slopes of the singular fields. In this case, the critical values of (K_σ, K_τ) of several islands with different size and shape show good agreement [Hirakata et al., 2006], and the toughness is evaluated as $(K_\sigma, K_\tau)_c$ = (0.24 MPam$^{0.39}$, 0.0038 MPam$^{0.60}$). This suggests that the local singular stress field on the order of 100 nm governs the crack initiation between the submicron island and the substrate. The fracture mechanics criterion is usually valid in the submicron scale.

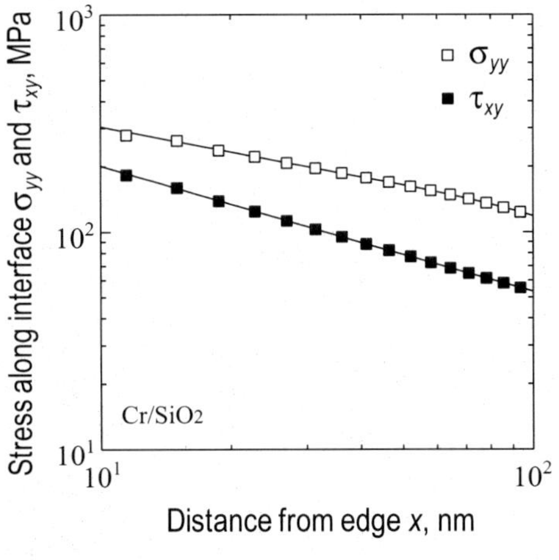

Figure 5.15 Stress distribution near the interface edge between the Cr island and SiO_2 substrate at delamination. Reprinted from Hirakata, H., et al. (2006) "Dominant stress region for crack initiation at interface edge of microdot on a substrate," *Eng. Fract. Mech.*, 73, pp. 2698–2709, Copyright (2006), with permission from Elsevier.

Strictly speaking, according to Bogy's equation [Bogy, 1968; Munz and Yang, 1993], there are two singularities, λ_A and λ_B, for the interface edge in this case so that the asymptotic stress field near the edge is expressed as the superposition of the two singularities.

$$\sigma_{ij}(r,\theta) = \frac{K_A}{r^{\lambda_A}} f_{ij}^{A}(\theta) + \frac{K_B}{r^{\lambda_B}} f_{ij}^{B}(\theta) \tag{5.3}$$

Here, $f_{ij}^{A}(\theta)$ and $f_{ij}^{B}(\theta)$ are the angular functions, which are 1 at $\theta = 0$ (on the interface). In this case, the stress intensity parameters K_A and K_B characterize the intensity of the singular stress field. The criterion is also expressed by $(K_A, K_B)_c$ = (0.029 MPam$^{0.50}$, 4.0 MPam$^{0.17}$). Because the parameters K_A and K_B do not directly correspond to the singularities of normal and shear stresses, a strength evaluation based on the parameters is not practical.

It should be noted that if there exists vertex of a nano-component on a substrate, as shown in Fig. 5.16, a delamination crack is likely to initiate at that point. This is because higher singularity appears for the three-dimensional vertex of bonded structures than of round-edged structures [Koguchi, 1997; Koguchi and Muramoto, 2000; Shibutani et al., 2003]. However, to the best of our knowledge, no precise experiments for the crack initiation at a vertex have been conducted on nano-scale components.

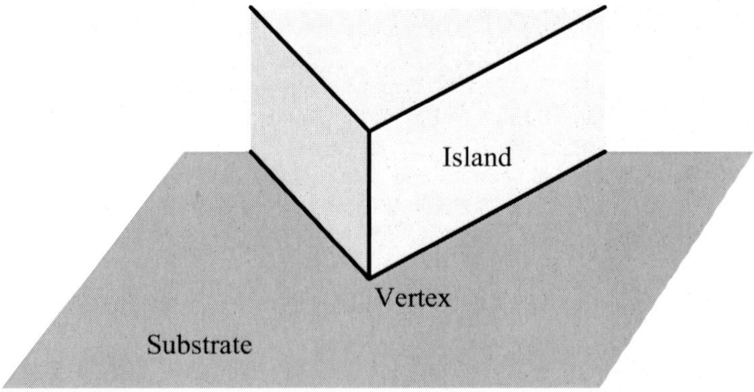

Figure 5.16 Schematic illustration of the vertex of an island on a substrate.

5.2.3 *In Situ* Observation of Crack Initiation and Stress Singularity in the Nano-scale

The method described in the previous section enables a delamination experiment for submicron components. But the local fracture process in the nano-scale cannot be observed in that method. In addition, it is difficult to scale down the specimen further and it cannot be modified for a tensile or bending test, which is a reliable loading method for mechanical investigation. A method designed to overcome these issues is a combination of *in situ* observation by transmission electron microscopy (TEM) and well-controlled nano-mechanical experiments where a small loading tip with a load sensor is equipped in the TEM holder [Hirakata et al., 2007]. The use of TEM allows us to observe not only the deformation and fracture processes of the specimen but also the dislocation motion and crystalline structures at the local region [Haque and Saif, 2005; Ohmura et al., 2005; Ii et al., 2004].

Figure 5.17 shows the testing setup, where a miniature mechanical loading apparatus, which consists of a movable sample stage and a diamond loading tip with a micro electromechanical system (MEMS) load sensor, is built into a TEM specimen holder. The stage, where the specimen and the gold wire (ϕ 0.25 mm) are mounted, can move three-dimensionally by means of a piezoelectric actuator. The alignment resolution of the actuator is approximately 1 nm for the x, y, and z axes. The load sensor detects the applied load by the capacitance change between the front and back electrodes, and is calibrated by a set of known weights. The effective load range for the sensor is 100 μN with a resolution of 0.1 μN.

Figure 5.18 shows the bend of the nano-cantilever consisting of SiN, Cu and Si for evaluating the crack initiation in nano-components [Hirakata et al., 2007]. The left and bottom ends of the Si plate are fixed, and a load is applied to the SiN layer by a loading tip so that the Cu/Si interface is stressed. A concentrated tensile stress is effectively applied to the interface edge near the fixed end of the cantilever by the bending moment. The effect of residual stress is negligible in this specimen because the normal stress on the interface dominates the cracking [Sumigawa et al., 2010b].

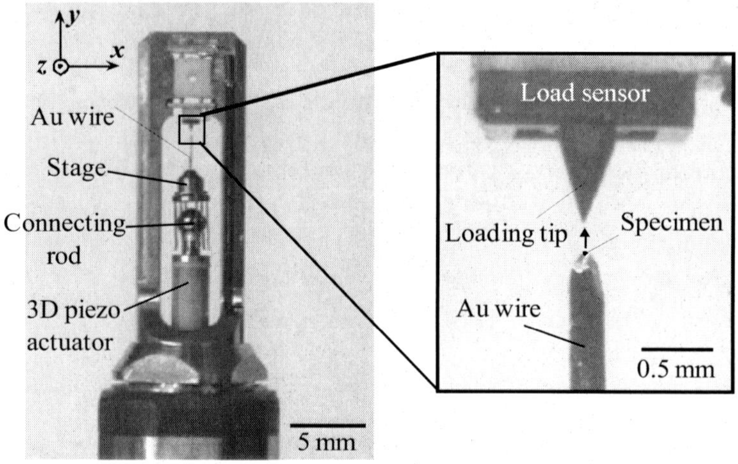

(a) Loading apparatus.

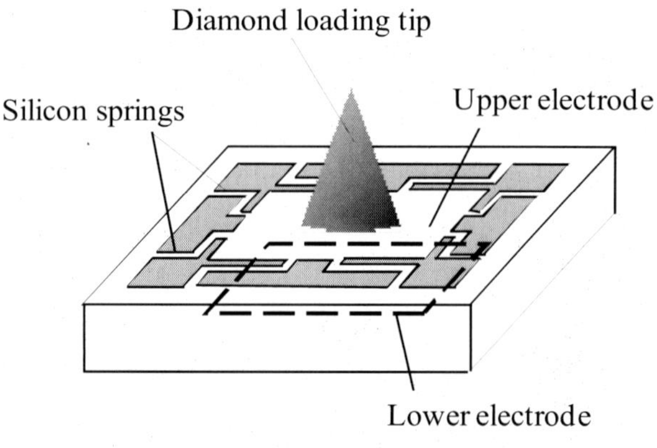

(b) MEMS load sensor.

Figure 5.17 Mechanical testing apparatus built into a TEM holder (Nanofactory Instruments AB, SA2000N). Reprinted from Hirakata, H., et al. (2007) "Role of plasticity on interface crack initiation at free edge and its propagation in nanostructures," *Int. J. Fract.*, 145, pp. 261–271, with permission from Springer.

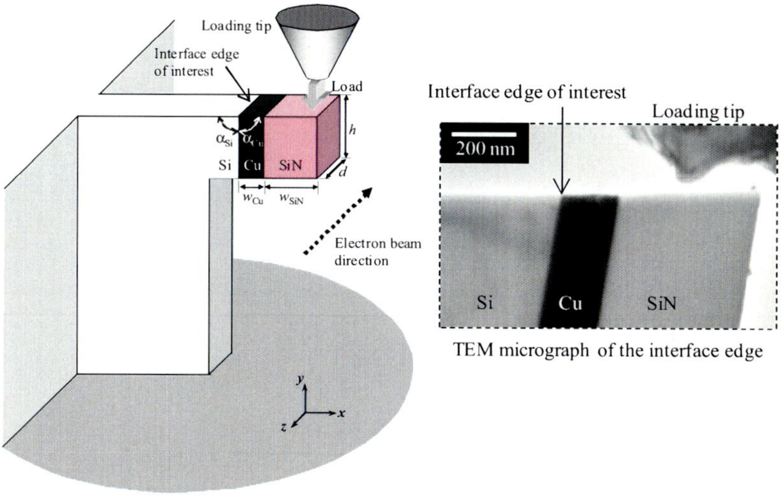

Figure 5.18 Crack initiation test at an interface edge in a nano-structure. Reprinted from Hirakata, H., et al. (2007) "Role of plasticity on interface crack initiation at free edge and its propagation in nanostructures," *Int. J. Fract.*, 145, pp. 261–271, with permission from Springer.

Figure 5.19 illustrates the specimen preparation procedure [Hirakata et al., 2007]. First, 500 nm-thick SiN and 200 nm-thick Cu films are deposited on a Si wafer by RF magnetron sputtering. The target cracking site is at the interface edge between the Cu and Si. A cubic block with side lengths of about 10 μm is cut from the multi-layered Si plate by means of a focused ion beam (FIB) (Fig. 5.19a). The block is picked up by a glass probe manipulator (Fig. 5.19b) and is carefully glued onto the top of a Au wire (Fig. 5.19c). After the block is thinned down to submicron scale in the *z* direction (Fig. 5.19d), a cantilever is formed by cutting a part of the membrane with the FIB (Fig. 5.19e). No damage such as crack and void is observed in the vicinity of the interface edge, as shown in the TEM micrograph, Fig. 5.18.

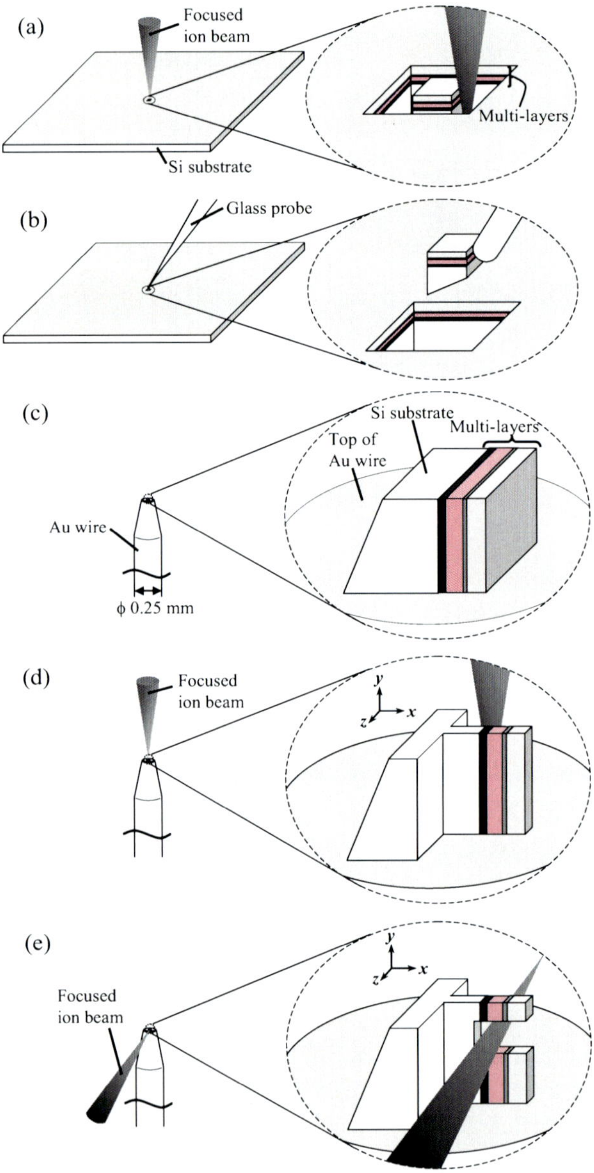

Figure 5.19 Preparation method of a nano-cantilever specimen. Reprinted from Hirakata, H., et al. (2007) "Role of plasticity on interface crack initiation at free edge and its propagation in nanostructures," *Int. J. Fract.*, 145, pp. 261–271, with permission from Springer.

Figure 5.20a,b shows the loading curve of a nano-cantilever specimen and *in situ* images of the specimen at three specific points, A, B, and C, in the loading curve. After the loading tip reaches the SiN layer at point A, the load increases monotonically up to the peak magnitude at point B. The load abruptly drops to zero (point C) at the critical load and the SiN and Cu layer delaminates from the Si plate. The magnified view of the Cu/Si interface edge shows no precursory crack formation at point B, which indicates that the crack is initiated immediately after point B and propagates in an instant along the Cu/Si interface.

Figure 5.20c shows the relationship between the applied load and the cantilever deflection. The deflection increases nonlinearly with the increase of the applied load. The observed nonlinear behavior of the load-deflection relation is attributed solely to the plastic deformation of the Cu component because the yield stresses of Si and SiN are much higher than that of Cu.

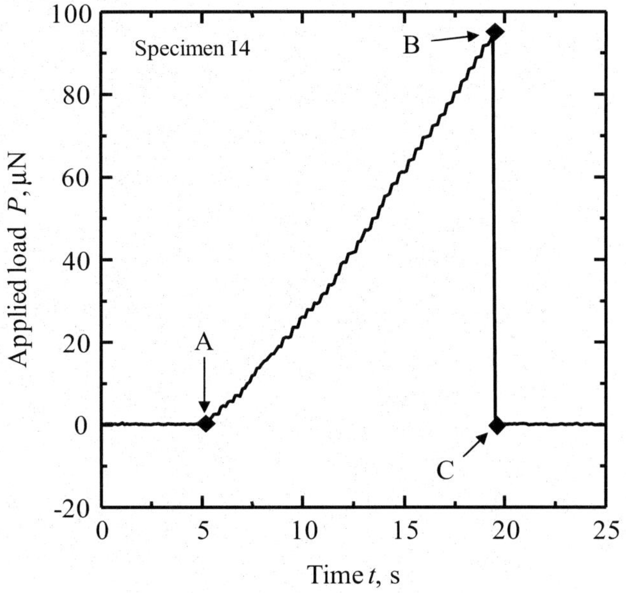

(a) Load-time relationship.

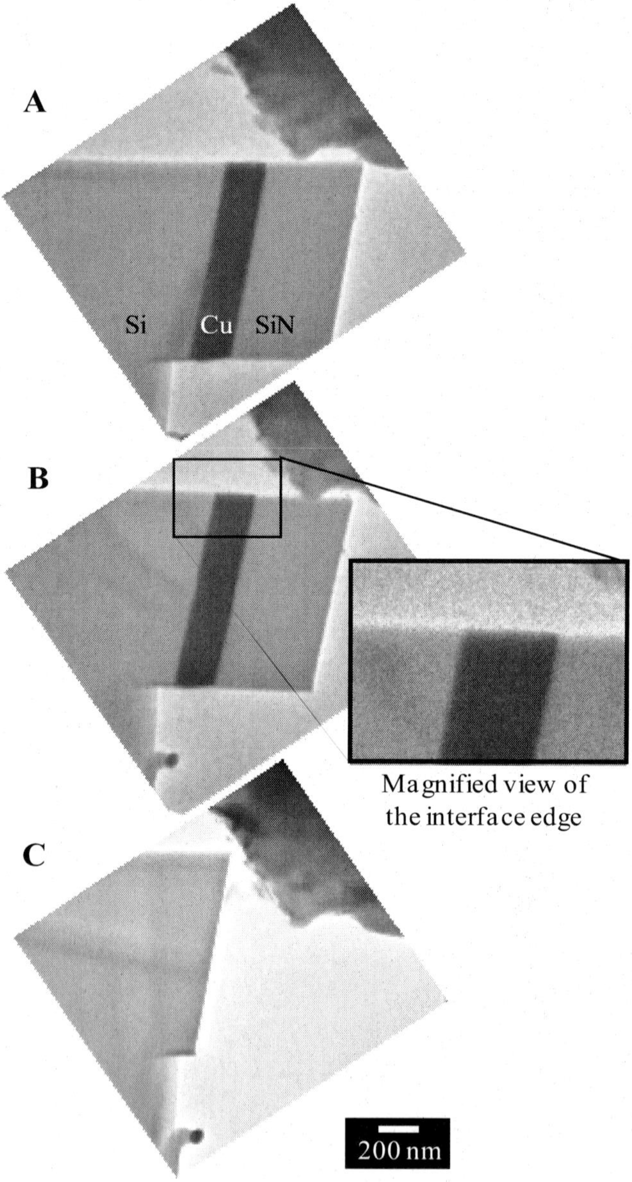

(b) Snapshots at points A, B, and C in Fig. 5.20a.

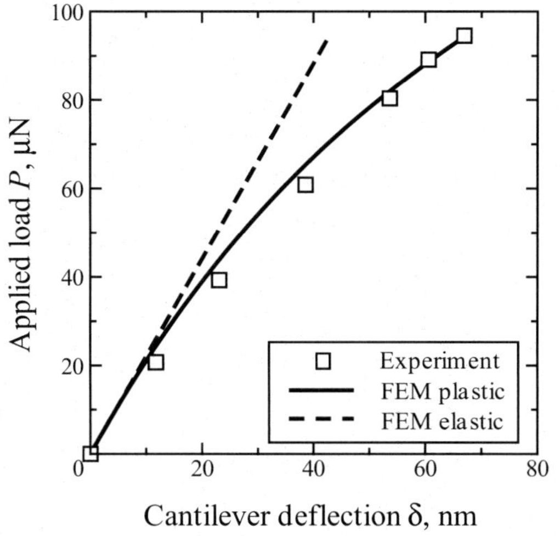

(c) Load- deflection relationship.

Figure 5.20 Relationships between load and time and between load and deflection in a crack initiation experiment using the nano-cantilever bend method. Reprinted from Hirakata, H., et al. (2007) "Role of plasticity on interface crack initiation at free edge and its propagation in nanostructures," *Int. J. Fract.*, 145, pp. 261–271, with permission from Springer.

To calculate precisely the stress distributions along the Cu/Si interface at crack initiation, the plasticity of Cu has to be considered. The plastic property of Cu is evaluated on the same material in the *in situ* TEM experiment [Hirakata et al., 2007; Takahashi et al., 2008]. The constants in the Ramberg-Osgood law

$$\bar{\varepsilon} = \frac{\bar{\sigma}}{E} + B\bar{\sigma}^n \tag{5.4}$$

are given as $B = 0.10\ \text{GPa}^{-n}$ and $n = 3.2$, respectively, by fitting the load-deflection curves in the nano-cantilever bend experiments with the results of elasto-plastic 3D-FEM analysis, as described in Section 3.4.4.1 We should be careful that the magnitude of n is different because Eq. (5.4) is different from Eq. (3.9).

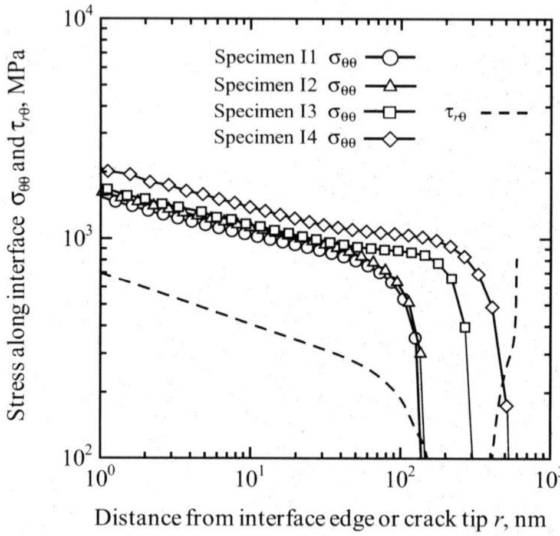

Figure 5.21 Elastic-plastic stress distributions near the interface edge between Si and Cu (200nm) at crack initiation. Reprinted from Hirakata, H., et al. (2007) "Role of plasticity on interface crack initiation at free edge and its propagation in nanostructures," *Int. J. Fract.*, 145, pp. 261–271, with permission from Springer.

Figure 5.21 shows the normal stress distributions considering Cu plasticity at crack initiation. The elastic-plastic singular stress field is expressed by an equation similar to that for the elastic singular field.

$$\sigma_{ij} = \frac{K_{ij}^{\mathrm{p}}}{r^{\lambda}} \tag{5.5}$$

Here, λ is about 0.160 for the Cu/Si interface edge, and the size of the singular stress field is about 40 nm. The stress intensity agrees well in the singular stress region for $r < 40$ nm among specimens of several sizes, while the stress distribution outside this region is very different. On the other hand, the intensity of the shear stress is lower than that of the normal stress, as shown by the dashed line in the figure. This signifies that normal stress dominates the delamination process. The crack initiation at the interface edge of the nano-components is dominated by the stress intensity K^{p}_{y}. The results have experimentally proved that fracture mechanics is still valid even on the order of 10 nm.

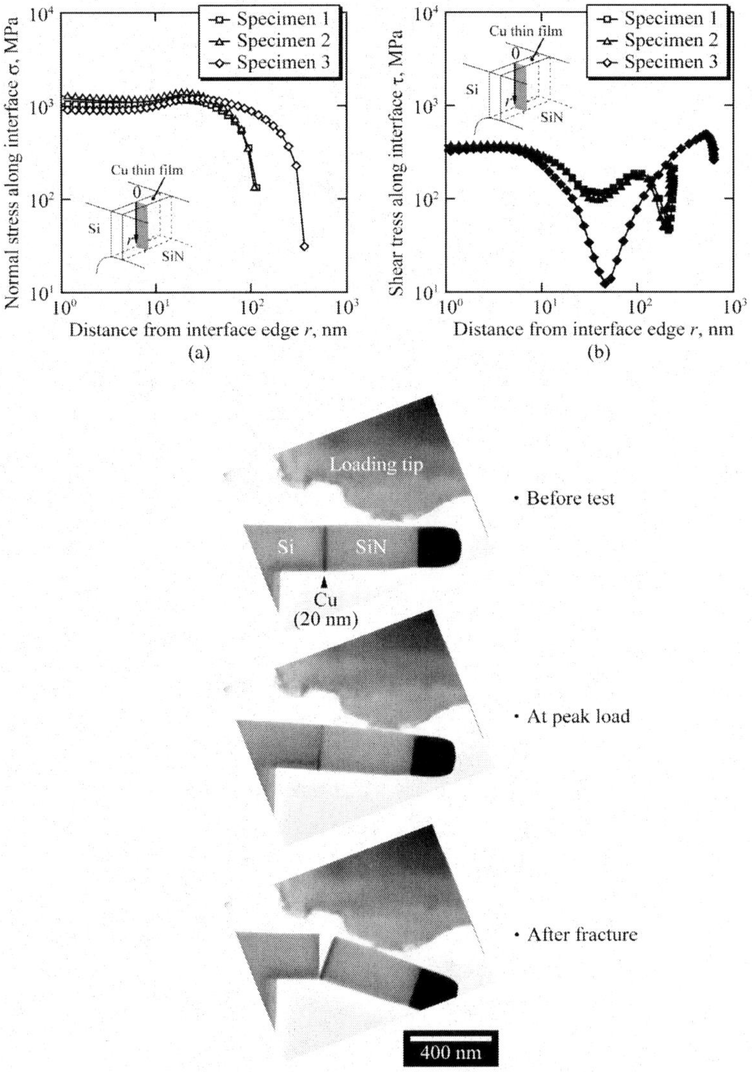

(c)

Figure 5.22 Elastic-plastic stress distributions near the interface edge between Si and Cu (20 nm) at crack initiation. (a) Normal stress. (b) Shear stress. (c) Fracture process. Reprinted from Sumigawa, T., et al. (2010) "Interface crack initiation due to nanoscale stress concentration," *Mater. Sci. Eng. A*, 527, pp. 4796–4803, Copyright (2010), with permission from Elsevier.

Figure 5.22 shows the result for specimens similar to the Cu film of 20 nm thickness [Sumigawa et al., 2010a]. The stress concentration near the interface edge governs the crack initiation.

Figure 5.23a shows a two-step cantilever specimen where the stress concentrates at the middle of the interface, as shown in Fig. 5.23b [Sumigawa et al., 2010a]. When the load is applied on the lower step, the crack initiation is found at the stress-concentrated point by *in situ* TEM observation. With the critical load experimentally measured, the stress for the interface crack initiation is examined by FEM, as plotted in Fig. 5.24. The cracking stress at the initiation site is about 1.1 GPa and the zone size over 1 GPa is about 60 nm. This corresponds fairly well with the one in the crack initiation at the interface edge shown in Fig. 5.22. This signifies that the interface crack initiation is brought about by a stress magnitude of about 1 GPa in the region of 50 nm, regardless of the location, suggesting that the criterion is universal for characteristic interface strength.

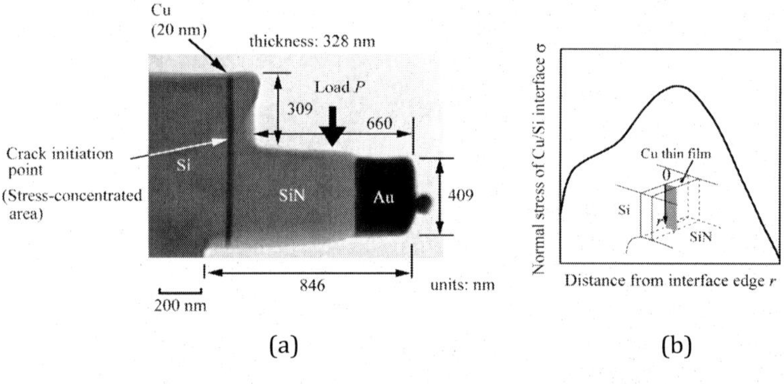

Figure 5.23 TEM image of two-step cantilever specimen (a) and the normal stress distribution along the Cu/Si interface (b). The material is the same as the one in Fig. 3.22. The thickness of the Cu is 20 nm. Reprinted from Sumigawa, T., et al. (2010) "Interface crack initiation due to nanoscale stress concentration," *Mater. Sci. Eng. A*, 527, pp. 4796–4803, Copyright (2010), with permission from Elsevier.

By changing the shape of the cantilever, as shown in Figs. 5.23a and 5.25, we can control the stress distribution along the interface. This may be useful for further explorations of the governing factors of cracking behavior in nano-components.

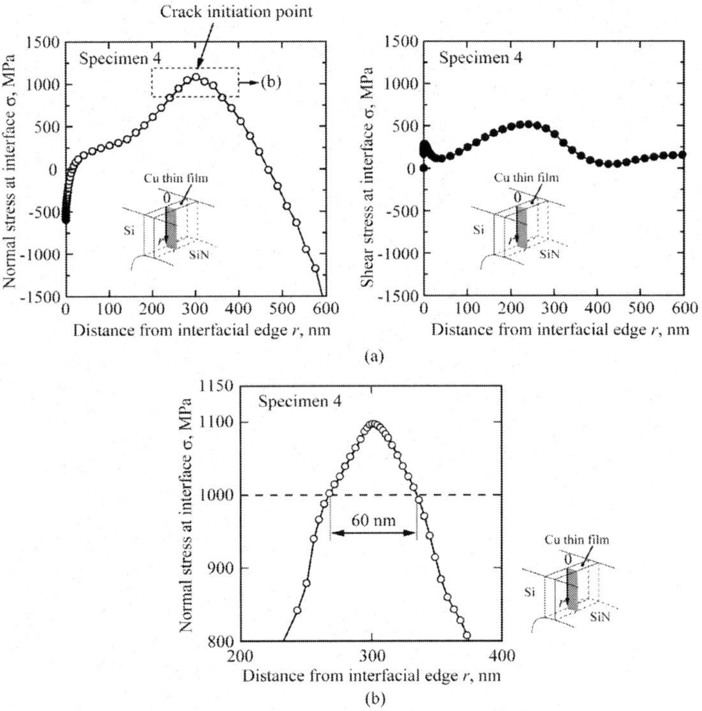

Figure 5.24 Stress distributions along the Cu/Si interface. The normal stress (left) and shear stress (right) are shown in (a). The normal stress at the concentrated area is magnified in (b). Reprinted from Sumigawa, T., et al. (2010) "Interface crack initiation due to nanoscale stress concentration," *Mater. Sci. Eng. A*, 527, pp. 4796–4803, Copyright (2010), with permission from Elsevier.

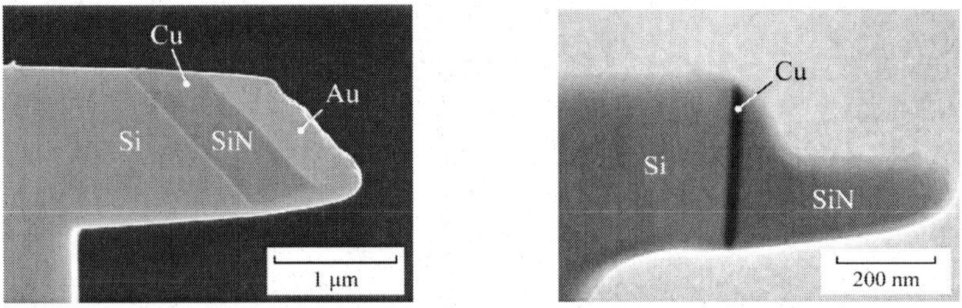

(a) (b)

Figure 5.25 Various specimens for the design of stress distribution along the interface. (a) Specimen with slant interface. (b) Specimen with slant surface at the interface edge.

5.2.4 Simplification of Cracking Criteria

Stress singularity at an interface edge λ depends not only on the material combination (modulus mismatch) but also on the corner shape of the interface edge. Hence, to apply the fracture mechanics concept to an interface when designing an industrial component, we need to evaluate experimentally the critical stress intensity for each λ of the corner shape. In addition, since the local stress state near the edges always becomes mixed-mode, even under a simple global loading mode, we have to consider the combination of the stress intensities for normal and shear stresses. Therefore, an enormous number of experiments with different edge shapes and loading modes are required to get a complete understanding of the cracking criteria of an interface. It is difficult in reality to conduct experiments that would cover all the possibilities, so a rational simplification is required. A possible candidate for simplified criterion might be an energy-based one corresponding to the energy release rate in the crack propagation. However, the energy release rate generally converges to zero as the crack length approaches zero. So, we cannot use the energy release rate as a fracture mechanics parameter for crack initiation at an interface edge. The validity of energy-based criteria is also questionable because energy dissipation occurs due to plastic deformation, as discussed in Section 5.2.3. A convenient numerical approach to overcome these problems is a cohesive zone model (CZM).

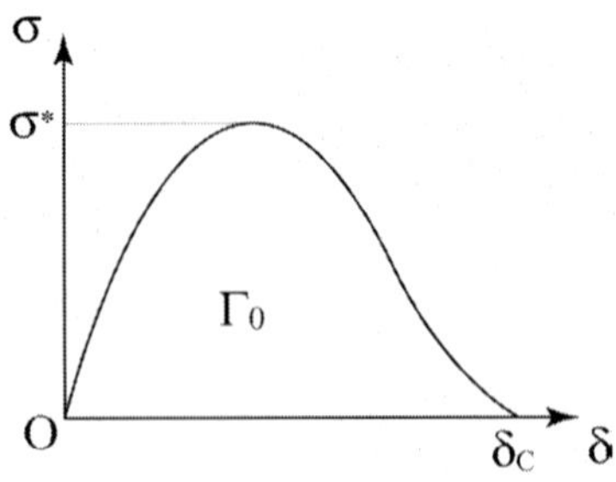

Figure 5.26 Constitutive relationship between traction and separation in a cohesive force model.

Let us consider a situation in which a crack initiates or propagates along an interface between dissimilar materials (materials 1 and 2) in a bonded structure. The so-called fracture process zone exists near the interface edge or the crack tip. In this region, the cohesive force between the materials σ is assumed to be a function of the separation of two materials δ. In the Barenblatt model [Barenblatt, 1962], σ represents the interatomic interaction, and the bond is cut off when δ reaches the critical length δ_C or the interaction becomes zero. The required energy for the complete separation Γ_0 is expressed as the following equation.

$$\Gamma_0 = \int_0^{\delta_C} \sigma(\delta)\mathrm{d}\delta \tag{5.6}$$

Since the size of structure is usually much larger than the interaction distance δ_C, which is on the order of 0.1 nm ~ 1 nm, the function of $\sigma(\delta)$ is not sensitive, and two parameters, namely the fracture energy Γ_0 and the maximum stress σ^*, characterize the fracture [Freund and Suresh, 2003]. The concept can be extended to complex fracture processes accompanied by large plastic deformation and micro-voids nucleation. In these cases, the detail of the local fracture is disregarded; instead the relationship between traction and displacement in the fracture process zone is assumed to obey the simple constitutive relationship $\sigma(\delta)$, as shown in Fig. 5.26.

CZMs based on the above concept are easily incorporated with FEM codes and can analyze a variety of fracture problems. For example, the effects of plastic deformation on interface crack propagation [e.g., Tvergaard and Hutchinson, 1994; Wei and Hutchinson, 1998; Lane et al., 2000], the mixed-mode criterion of interface cracks [Kishimoto et al., 2002], and so on have been addressed.

A unified (approximate or simplified) criterion based on a CZM is discussed for the crack initiation from interface edges having different edge shapes in bulk bi-materials [Mohammed and Liechti, 2000] and in submicron films [Zhao et al., 2010].

An approach using CZM may enable us to evaluate *ad hoc* criteria for the interface crack initiation of nano-scale components [Yan and Shang, 2009].

The method, however, is based on simple modeling although the fracture process at interfaces in atomic level is complex. Hence, the rationale of the physical process is ambiguous. In order to elucidate reasonably the fracture criterion on the nano-scale where continuum mechanics breaks down, it is necessary to use discrete dislocation mechanics (DD) [O'Day et al., 2006; Chng et al., 2006], molecular dynamics (MD) [Kitamura et al., 2004], and/or first principle calculation. These approaches are discussed in Chapters 7 and 8.

5.3 CREEP

Metals with a low melting temperature, e.g. solder, and polymers are widely used in micro-electronic devices. Some of these materials are exposed to elevated temperatures in the manufacturing process or in service, so creep deformation takes place even at low stress levels. Thus, fracture (crack initiation) due to creep has been studied as an important issue in the industry. In particular, the characteristic feature in creep is its time-dependent behavior, where the damage gradually accumulates and leads to a complete fracture. Although creep crack initiation at a notch root in homogeneous materials has been studied well [e.g., Ohtani, 1979], creep crack initiation from an interface edge has not yet been investigated yet in terms of fracture mechanics, even for bulk materials [Kitamura et al., 2007b; Ngampungpis et al., 2007, 2008].

Figure 5.27 shows a cantilever specimen for a study of creep crack initiation from an interface edge at a room temperature [Hirakata et al., 2008]. The cantilever consists of Si, Sn, and Ta_2O_5 layers and is formed on a part of the substrate by a FIB. The interface of interest is the one between the Sn and Si. Sn has relatively a low melting temperature of 505 K so creep deformation becomes significant even at room temperature, while creep hardly ever occurs in Si and Ta_2O_5. As shown in Fig. 5.27a, a load is applied using testing apparatus similar to that shown in Section 5.2.2

(Fig. 5.8b) to the top surface of the Ta_2O_5 layer by a tetrahedral diamond tip with a vertical angle of 60°.

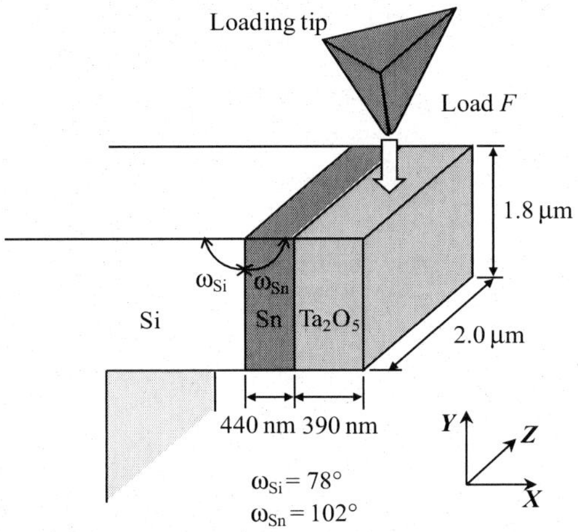

(a) Schematic illustration of a specimen and loading method.

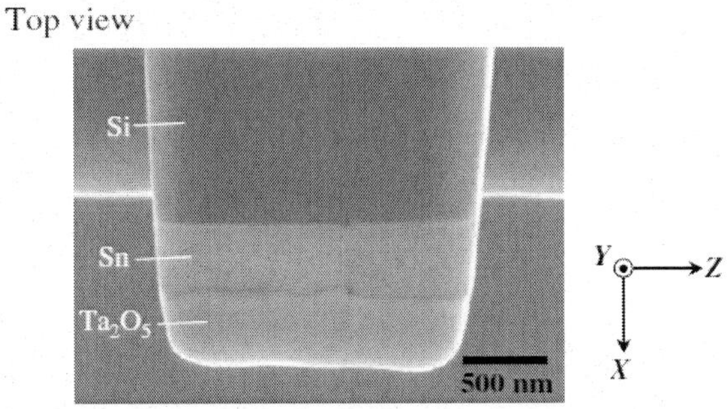

(b) FE-SEM micrograph of a tested specimen.

Figure 5.27 Micro-cantilever bend specimen for crack initiation experiments due to creep. Reprinted from Hirakata, H., et al. (2008) "Creep crack initiation at a free edge of an interface between submicron thick elements," *Eng. Fract. Mech.*, 75, pp. 2907–2920, Copyright (2008), with permission from Elsevier.

Figure 5.28a shows the change in the displacement at the loading point over time. After the initial loading up to point "a", the displacement increases to "b" with the passage of time under the constant load. In the transient creep region, the creep decelerates, in the early period up to 200 s. The creep rate becomes almost constant after that: i.e., it becomes steady state creep. At point "b", the displacement abruptly increases and the Sn layer completely delaminates from the Si substrate, as shown in the field emission scanning electron microscopy (FE-SEM) image in Fig. 5.28b. The crack clearly goes through the specimen along the interface between the Sn and Si, and the dent on the Ta_2O_5 layer confirms that the load is applied to the intended layer.

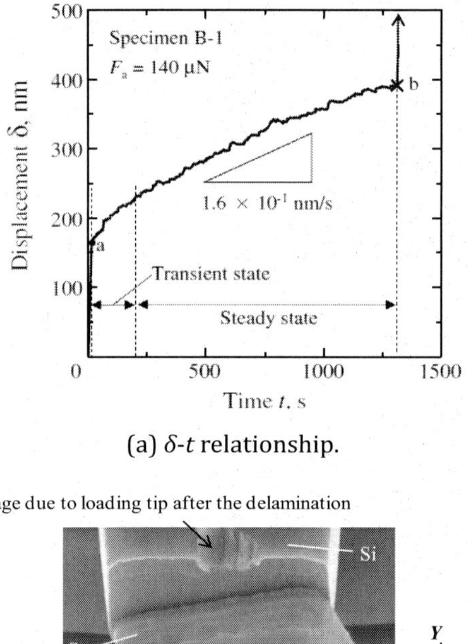

(a) δ-t relationship.

(b) FE-SEM micrograph of a specimen after the experiment.

Figure 5.28 Test results of a time-dependent crack initiation experiment under a constant load. Reprinted from Hirakata, H., et al. (2008) "Creep crack initiation at a free edge of an interface between submicron thick elements," *Eng. Fract. Mech.*, 75, pp. 2907–2920, Copyright (2008), with permission from Elsevier.

The creep property (Norton's power law) of Sn under steady state creep is

$$\dot{\bar{\varepsilon}} = 1.9 \times 10^{-10} \bar{\sigma}^{3.0} \tag{5.7}$$

where $\bar{\sigma}$ and $\dot{\bar{\varepsilon}}$ are the von Mises equivalent stress and strain rate, which are inversely evaluated from experiments and FEM analysis (see Section 3.4.3.2). Figure 5.29 shows the transition of von Mises stress distribution along the interface near the edge. Just after the loading instant t = 0.1 s, where the creep region is very small, the elastic stress singularity of λ_s^{el} = 0.028 is dominant. The creep region quickly extends with the passage of time, and the stress field reaches a steady state at about t = 17 s. After that, the stress field does not change up to $t \approx 1300$ s. The stress singularity in the creep region λ_s^{cr} is 0.23 including the transition period; Fig. 5.30 shows the change in the stress intensity. Although the stress intensity increases in the very early region, it is in the steady state in almost all of the period. So, the effect of the transient state on the creep crack initiation behavior is negligible, although the transient state should be considered for strict evaluation. Figure 5.31 shows the relationship between von Mises stress intensity at the steady state $K_{s\ \text{m-steady}}^{cr}$ and crack initiation time t_C. The experiments show excellent repeatability, implying that the crack initiation from the edge is governed by the stress intensity at the interface edge, i.e., $K_{s\ \text{m-steady}}^{cr}$.

$$K_{s\ \text{m-steady}}^{cr} \cdot t_C^{0.14} = 10.6\ \text{MPa m}^{0.23}\text{s}^{0.14} \tag{5.8}$$

These facts suggest that fracture mechanics can be applied to creep crack initiation (time-dependent fracture).

It is well known that the mechanisms of crack initiation in bulk materials strongly depend on conditions such as fatigue (cyclic loading), environment (mechano-chemical fracture), and creep. The results in this section indicate that, even for nano-materials, the fracture mechanisms depend on external factors, and the characteristics and nature of a crack initiation can vary in the extreme. It also suggests that fracture mechanics is still valid at the submicron scale, even for complex fractures such as creep. Considering the results of toughness laid out in Section 5.2, the mechanics must be applicable at a scale of 10 nm, although an experiment proving this has not yet been conducted. On the other hand, the effect of environment (chemical factors) will be enhanced as the size of the component is scaled down, but no critical experiments showing this have been conducted yet either. In addition, exploring the fracture mechanics of fatigue cracking in small components, where the local region of the micron- to nanometer-scale dominates, will be valuable.

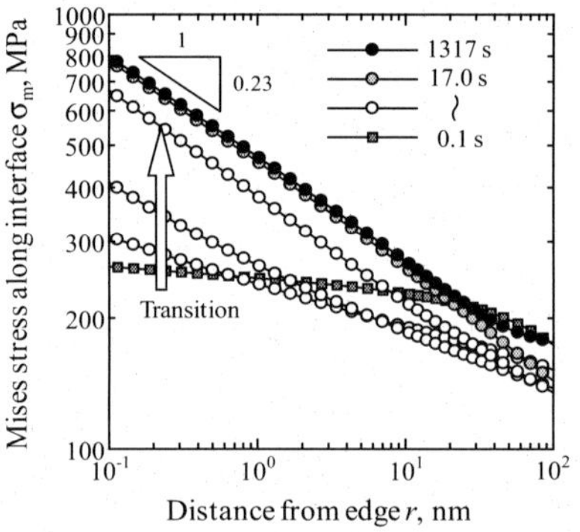

Figure 5.29 Change in Mises stress distribution near the Sn/Si interface edge during the crack initiation experiment; a singular stress field exists and the intensity time-dependently increases. Reprinted from Hirakata, H., et al. (2008) "Creep crack initiation at a free edge of an interface between submicron thick elements," *Eng. Fract. Mech.*, 75, pp. 2907–2920, Copyright (2008), with permission from Elsevier.

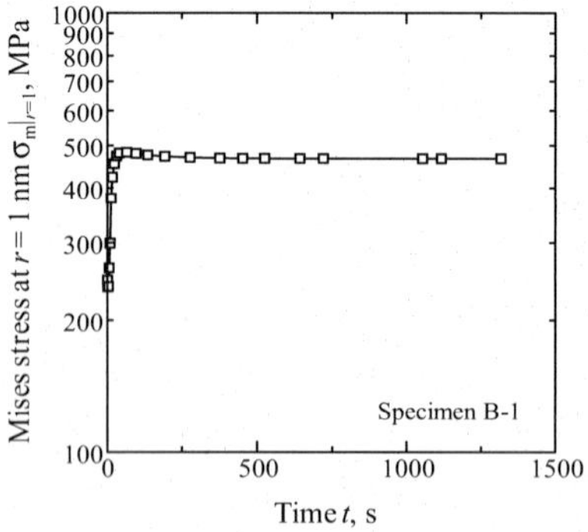

Figure 5.30 Change in von Mises stress at $r = 1$ nm from the Sn/Si interface edge; the stress increases at the early stage, then reaches a steady state. Reprinted from Hirakata, H., et al. (2008) "Creep crack initiation at a free edge of an interface between submicron thick elements," *Eng. Fract. Mech.*, 75, pp. 2907–2920, Copyright (2008), with permission from Elsevier.

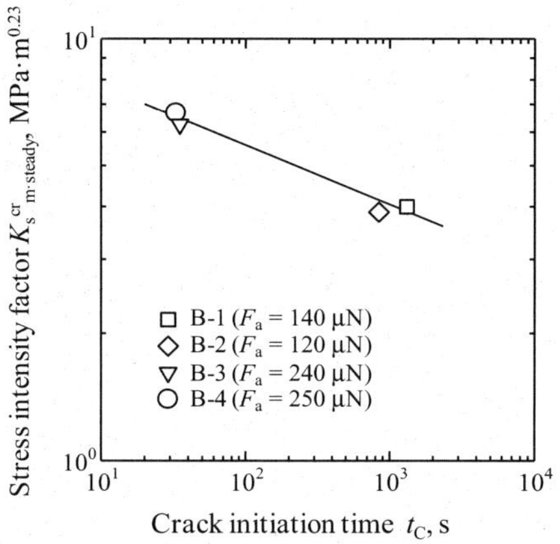

Figure 5.31 Relationship between creep stress intensity factor at the steady state and crack initiation time. Reprinted from Hirakata, H., et al. (2008) "Creep crack initiation at a free edge of an interface between submicron thick elements," *Eng. Fract. Mech.*, 75, pp. 2907–2920, Copyright (2008), with permission from Elsevier.

5.4 FATIGUE

The cantilever specimen with an interface between a 20 nm-thick Cu and a Si substrate is subjected to a cyclic load with the range of 18 μN ($\Delta P = P_{max} - P_{min}$) and a load ratio of $P_{min}/P_{max} = 0$ in a vacuum of 1.5×10^{-5} Pa, as illustrated in Fig. 5.32 (P_{min} is minimum load; P_{max} is maximum load). The fatigue process and the displacement at the arm end, δ, are observed *in situ* by means of the TEM. The cyclic tensile strain is concentrated in the region of about 40 nm in the Cu layer and on the Si/Cu interface from the top surface [Sumigawa et al., 2010c].

In the last cycle, the load suddenly drops at 17.0 μN before the maximum load and the Si/Cu interface breaks. The failure load is about 6% smaller than the maximum load in the cycle and is much lower than the expected fracture load in monotonic loading.

The P-δ curve in a cycle (Fig. 5.32b) shows nonlinear behavior and a distinct hysteresis loop, which indicates the cyclic plasticity of the Cu film because the Si substrate and the SiN layer are elastically deformed at this load level. The cyclic hardening is reported from the change in the hysteresis. These suggest the development of a dislocation substructure in the Cu film.

In the fatigue of bulk metals it is well known that irreversible cyclic deformation brings about characteristic dislocation structures such as a vein, a ladder structure, or a cell. In particular, the ladder structure and cell cause crack initiation owing to strain localization because they are much softer than the matrix. However, these substructures have a size of a few micrometers. Since the strain concentrated area (about 40 nm × 20 nm × 330 nm in this specimen) is on a nanometer scale, these are not developed in the Cu layer. Because our knowledge of fatigue in nano-components is very limited, the peculiar features of dislocation substructures are totally unknown, but further investigation is inevitable.

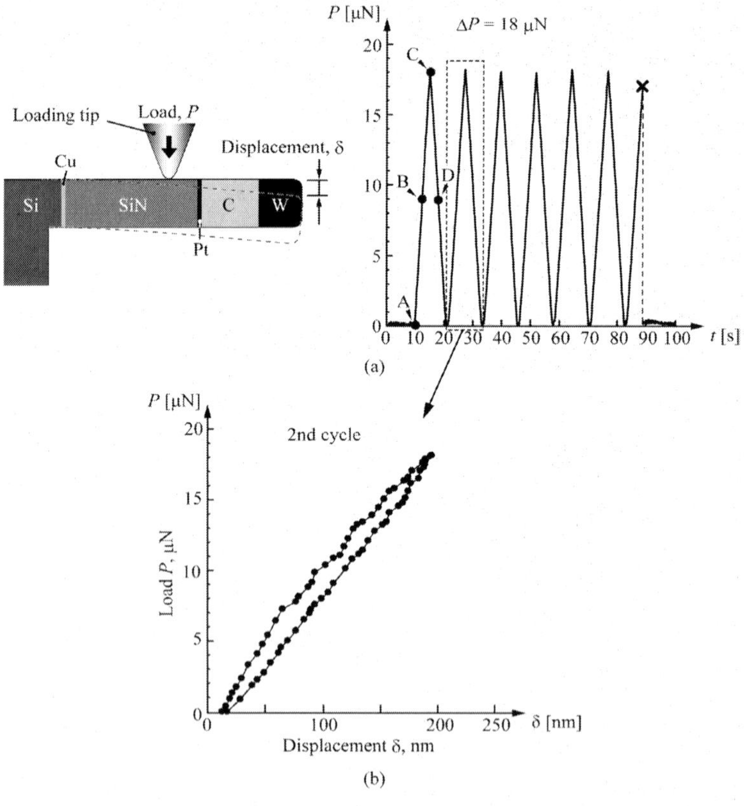

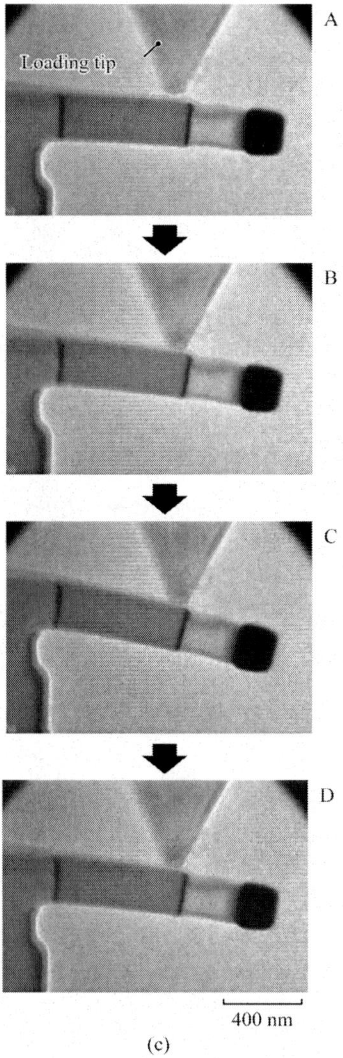

Figure 5.32 Cantilever specimen for fatigue experiment and cyclic loading cycle where X indicates the failure along the Si/Cu interface. The irreversible deformation (b) and (c) is observed in a fatigue cycle (a) . Reprinted from Sumigawa, T., et al. (2010) "Cu/Si interface fracture due to fatigue of copper film in nanometer scale," *Mater. Sci. Eng. A*, 527, pp. 4796–4803, Copyright (2010), with permission from Elsevier.

References

Barenblatt, G. I. (1962) The mechanical theory of equilibrium cracks in brittle materials, *Adv. Appl. Mech.*, 7, pp. 55–129.

Becker Jr., T. L., McNaney, J. M., Cannon, R. M., and Ritchie, R. O. (1997) Limitations on the use of the mixed-mode delamination beam test specimen: effects of the size of the region of K-dominance, *Mech. Mater.*, 25, pp. 291–308.

Bogy, D. B. (1968) Edge bonded dissimilar orthogonal elastic wedges under normal and shear stress, *J. Appl. Mech.*, 35, pp. 146–154.

Chng, A. C., O'Day, M. P., Curtin, W. A., Tay, A. A. O., and Lim, K. M. (2006) Fracture in confined thin films: a discrete dislocation study, *Acta Mater.*, 54, pp. 1017–1027.

Freund, L. B., and Suresh, S. (2003) *Thin Film Materials: Stress, Defect Formation and Surface Evolution*, Cambridge University Press.

Haque, M. A., and Saif, M. T. A. (2005) *In situ* tensile testing of nanoscale freestanding thin films inside a transmission electron microscope, *J. Mater Res.*, 20, pp. 1769–1777.

Hirakata, H., Kitamura T., and Yamamoto Y. (2004a) Evaluation of interface strength of micro-dot on substrate by means of AFM, *Int. J. Solids Struct.*, 41, pp. 3243–3253.

Hirakata, H., Kitamura, T., and Yamamoto, Y. (2004b) Direct measurement of interface strength between copper submicron-dot and silicon dioxide substrate, *JSME Int. J. Ser. A*, 47, pp. 324–330.

Hirakata, H., Takahashi, Y., Matsumoto, S., and Kitamura, T. (2006) Dominant stress region for crack initiation at interface edge of microdot on a substrate, *Eng. Fract. Mech.*, 73, pp. 2698–2709.

Hirakata, H., Takahashi, Y., Truong, D. V., and Kitamura, T. (2007) Role of plasticity on interface crack initiation at free edge and its propagation in nanostructures, *Int. J. Fract.*, 145, pp. 261–271.

Hirakata, H., Hirako, T., Takahashi, Y., Matsuoka, Y., and Kitamura, T. (2008) Creep crack initiation at a free edge of an interface between submicron thick elements, *Eng. Fract. Mech.*, 75, pp. 2907–2920.

Ii, S., Iwamoto, C., Matsunaga, K., Yamamoto, T., and Ikuhara, Y. (2004) Identification of crack path of inter- and transgranular fractures in sintered silicon nitride by *in situ* TEM, *J. Electron Microscopy*, 53, pp.121–127.

Kishimoto, K., Omiya, M., and Yang, W. (2002) Fracture mechanics of bonding interface: a cohesive zone model, *Sensors and Actuators A*, 99, pp. 198–206.

Kiramura, T., Shibutani, T., and Ueno, T. (2002) Crack initiation at free edge of interface between thin films in advanced LSI, *Eng. Fract. Mech.*, 69, pp. 1289–1299.

Kitamura, T., Hirakata H., and Itsuji T. (2003) Effect of residual stress on delamination from interface edge between nano-films, *Eng. Fract. Mech.*, 70, pp. 2089–2101.

Kitamura, T., Hirakata, H., and Satake, Y. (2004) Applicability of fracture mechanics on brittle delamination of nanoscale film edge, *JSME Int. J. Ser. A*, 47, pp. 106–112.

Kitamura, T., Hirakata, H., and Truong, D. V. (2007a) Initiation of interface crack at free edge between thin films with weak stress singularity, *Thin Solid Films*, 515, pp. 3005–3010.

Kitamura, T., Ngampungpis, K., and Hirakata, H. (2007b) Stress field near interface edge of elastic-creep bi-material, *Eng. Fract. Mech.*, 74, pp. 1637–1648.

Koguchi, H. (1997) Stress singularity analysis in three-dimensional bonded structure. *Int. J. Solids Struct.*, 34, pp. 461–480.

Koguchi, H., and Muramoto, T. (2000) The order of stress singularity near the vertex in three-dimensional joints, *Int. J. Solids Struct.*, 37, pp. 4737–4762.

Lane, M., Dauskardt, R. H., Vainchtein, A., and Gao, H. (2000) Plasticity contributions to interface adhesion in thin-film interconnect structures, *J. Mater. Res.*, 15, pp. 2758–2769.

Mohammed, I., and Liechti, M. (2000) Cohesive zone modeling of crack initiation at biomaterial corners, *J. Mech. Phys. Solids*, 48, pp. 735–764.

Munz, D., and Yang, Y. Y. (1993) Stresses near the edge of bonded dissimilar materials described by two stress intensity factors, *Int. J. Fract.*, 60, pp. 169–177.

Ngampungpis, K., Kitamura, T., and Hirakata, H. (2007) Effect of material thickness on the singular stress field in elastic-creep bi-material interface, *J. Solid Mech. Mater. Eng.*, 1, pp. 744–754.

Ngampungpis, K., Kitamura, T., and Hirakata, H. (2008) Increase of stress intensity near interface edge of elastic-creep bi-material under a sustained load, *Eng. Fract. Mech.*, 75, pp. 1285–1293.

O'Day, M. P., Nath, P., and Curtin, W. A. (2006) Thin film delamination: a discrete dislocation analysis, *J. Mech. Phys. Solids*, 54, pp. 2214–2234.

Ohmura, T., Minor, A. M., Stach, E. A., and Morris Jr., J. W. (2005) Dislocation-grain boundary interactions in martensitic steel observed through in situ nanoindentaion in a transmission electron microscope, *J. Mater. Res.*, 19, pp. 3626–3632.

Ohtani, R. (1979) Stress and strain concentrations of notched members in creep, *J. Soci. Mater. Sci. Japan*, 25, pp. 230–235 (in Japanese).

Shang, F., Kitamura, T., Hirakata H., Kanno, I., Kotera, H., and Terada, T. (2005a) Experimental and theoretical investigations of delamination at free edge of interface between piezoelectric thin films on a substrate, *Int. J. Solids Struct.*, 42, pp. 1729–1741.

Shang, F., Kitamura, T., and Hirakata, H. (2005b) Delamination test and the effect of free edge on interface strength of PZT thin films, *Integrated Ferroelectrics*, 73, pp. 67–74.

Shibutani, T., Tsuruga, T., Yu, Q., and Shiratori, M. (2003) Evaluation of interface strength between thin films fabricated on a silicon substrate for an advanced LSI on the basis of fracture mechanics concept, *Mater. Sci. Forum*, 426–432, pp. 3469–3474.

Sumigawa, T., Shishido, T., Murakami, T., and Kitamura, T. (2010a) Interface crack initiation due to nano-scale stress concentration, *Mater. Sci. Eng. A*, 527, pp. 4796–4803.

Sumigawa, T., Shishido, T., Murakami, T., and Kitamura, T. (2010b) Evaluation on plastic deformation property of copper nano-film by nano-scale cantilever specimen, *Thin Solid Films*, 518, pp. 6040–6047.

Sumigawa, T., Murakami, T., Shishido, T., and Kitamura, T. (2010c) Cu/Si interface fracture due to fatigue of copper film in nanometer scale, *Mater. Sci. Eng. A*, 527, pp. 6518-6523.

Takahashi, Y., Hirakata, H., and Kitamura, T. (2008) Quantitative evaluation of plasticity of a ductile nano-component, *Thin Solid Films*, 516, pp. 1925–1930.

Tvergaard, V., and Hutchinson, J. W. (1994) Toughness of an interface along a thin ductile layer joining elastic solids, *Philos. Mag. A*, 70, pp. 641–656.

Wei, Y., and Hutchinson, J. W. (1998) Interface strength, work of adhesion and plasticity in the peel test, *Int. J. Fract.*, 93, pp. 315–333.

Yan, Y., and Shang, F. (2009) Cohensive zone modeling of interfacial delamination in PZT thin film, *Int. J. Solids and Struct.*, 46, pp. 2739–2749.

Zhao, P., Shang, F., Yan, Y., and Kitamura, T. (2010) Reinvestigation of the effect of plasticity on delamination process of Si/Cu interface in a ductile nano-cantilever, *Int. J. Fract.*, 161, pp. 65–78.

Chapter 6

COMPONENTS CONSISTING OF NANO-ELEMENTS

6.1 STRUCTURES

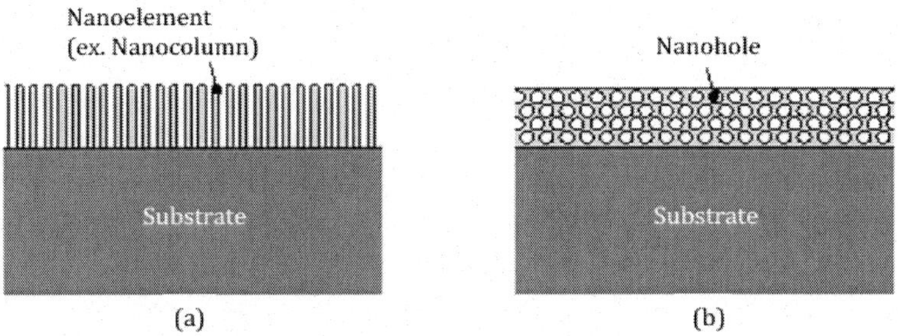

Figure 6.1 Thin films (a) composed of aggregated nano-elements and (b) with dispersed nano-holes.

With the development of process technology, it has become possible to fabricate thin films comprised of aggregated nanometer-scale elements (nano-elements), as illustrated in Fig. 6.1a [Hersee et al., 2006; Kipshidze et al., 2005; Mohan et al., 2005; Tomioka et al., 2007; Dyke and Martin, 1990; Lee et al., 2007; Lai et al., 2008]. Because of its characteristic structure, the film is expected to have exotic mechanical properties. A porous film with tunable nano-holes, illustrated in Fig. 6.1b [Guo et al., 2005; Chung et al., 2007; Jin et al., 1998; Zhu and Ding, 1999; Kohl et al., 1999; Shin et al., 2005;

Fracture Nanomechanics
Edited by Takayuki Kitamura, Hiroyuki Hirakata, Takashi Sumigawa and Takahiro Shimada

www.panstanford.com

Ekanayake et al., 2007; Zang et al., 2005], is in this family in terms of its mechanical behavior. Although we discuss only the examples of nano-elements in Fig. 6.1a in this chapter, that behavior is easily extended to films with nano-holes, shown in Fig. 6.1b.

There are methods for fabricating aggregated film for various materials. Typical ones are as follows.

6.1.1 Dynamic Oblique Deposition (DOD) Method

In the DOD method [Robbie et al., 1995; Robbie and Brett, 1997; Suzuki and Taga, 2001a; Malac and Egerton, 2001], the atoms physically vapored are deposited on a substrate at an oblique angle. Once the nuclei have formed on the substrate, a self-shadowing effect makes the elements grow in a certain direction. Various shapes (e.g., helical, zigzag, pillars) can be fabricated by adjusting the rotation of the substrate and the angle of incident vapor flux. This is explained in greater detail in Section 6.2.1.

6.1.2 DC Plasma-Enhanced Chemical Vapor Deposition (PECVD)

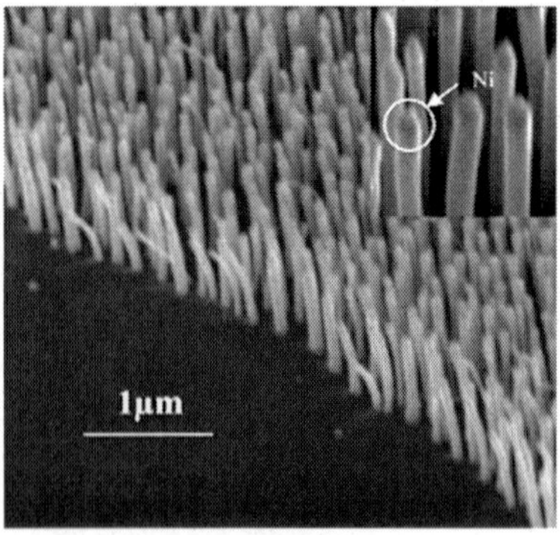

Figure 6.2 Carbon nanotubes (CNTs) vertically grown on a silicon substrate by patterning Ni seed layer. Reprinted with permission from Abdi, Y., et al. (2006) "Carbon nanostructures on silicon substrates suitable for nanolithography," *Appl. Phys. Lett.*, 88, article number 053124. Copyright (2006), American Institute of Physics.

PECVD [Abdi et al., 2006; Kato et al., 2004; Meyyappan et al., 2003; Qu et al., 2008; Melechko et al., 2005] is a method for fabricating a forest of multi-wall carbon nanotubes on a substrate. After particles of catalytic metal (e.g., Fe, Co, and Ni) are formed on a substrate, carbon nanotubes (CNTs) are perpendicularly grown from them by the thermal decomposition of carbon hydride gas. Massive production is available by this method. Figure 6.2 shows the carbon nanotube arrays obtained by PECVD.

6.1.3 Hydrothermal Crystallization of Colloidal Precursors

Colloidal precursors formed on a substrate are crystallized in high-temperature aqueous solution [Cao et al., 2007; Elias et al., 2008]. Anisotropic nanostructures, of a scale from a few tens of nanometers to a few microns, are grown from them. This process is easier than the direct crystallization of nano-particles from solution because the energy barrier in heterogeneous nucleation is smaller than that in homogeneous nucleation. This process can form nanostructures extended in various directions according to the crystallographic orientation. Figure 6.3 shows ZnO nanowires obtained by this technique.

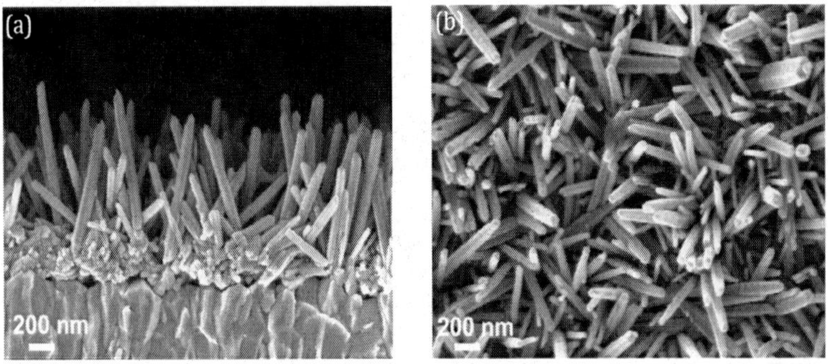

Figure 6.3 ZnO nanowire arrays electrodeposited on a substrate. The nanowires with a rod shape extend in different directions according to the crystallographic orientation. (a) Side view. (b) Top view. Reprinted from Elias, J., Tena-Zaera, R., and Levy-Clement, C. (2008) "Electrochemical deposition of ZnO nanowire arrays with tailored dimensions," *J. Electroanal. Chem.*, 621, pp. 171 -177, Copyright (2008), with permission from Elsevier.

6.1.4 Template-Synthesis Method

In the template-synthesis method [Yao et al., 2002; Chu et al., 2003; Vazquez et al., 2006; Lee et al., 2005; Hong et al., 2007], a polymer template is molded using a (silicon) substrate with a nano-scale pattern formed by etching (e.g., deep reactive ion etching). After the template is removed from the substrate, a thin film is deposited on the template by liquid phase infiltration (deposition in aqueous solution using ligand-exchange (hydrolysis) equilibrium reaction) or by evaporation in vacuum. When the template is dissolved, a film with nano-pattern is obtained. Template-synthesis makes it easy to control the shape and the space between elements, and makes it possible to fabricate complex morphologies, e.g., glass wool or porous materials. Figure 6.4 shows Ni nanowire arrays standing on a substrate, where each wire has nearly same diameter.

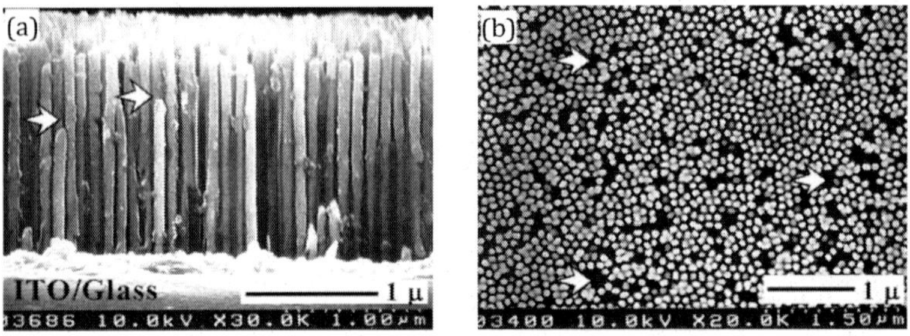

Figure 6.4 Ni nanowire arrays deposited on a glass substrate coated with a tin-doped indium oxide film. (a) Side view. (b) Top view. Reprinted from Chu, S. Z., et al. (2003) "Fabrication and characteristics of nanostructures on glass by Al anodization and electrodeposition," *Electrochim. Acta*, 48, pp. 3147–3153, Copyright (2003), with permission from Elsevier.

6.1.5 Molecular Beam Epitaxy (MBE) Method

The molecular beam epitaxy (MBE) technique [Sanchez-Paramo et al., 2002; Kim et al., 2005; Wang et al., 2006; Bougrioua et al., 2007], performed in a high vacuum condition (≤10^{-8}Pa), enables us to make the deposition slowly, which brings about epitaxial growth. This also makes it possible to fabricate film of various chemical compounds. By controlling the deposition angle, we can extend nano-elements in a given direction from the nuclei formed on a substrate. It is rather easy to control the chemical compound in the elements. Figure 6.5 shows ZnO nanorod arrays formed on a silicon substrate using the MBE method.

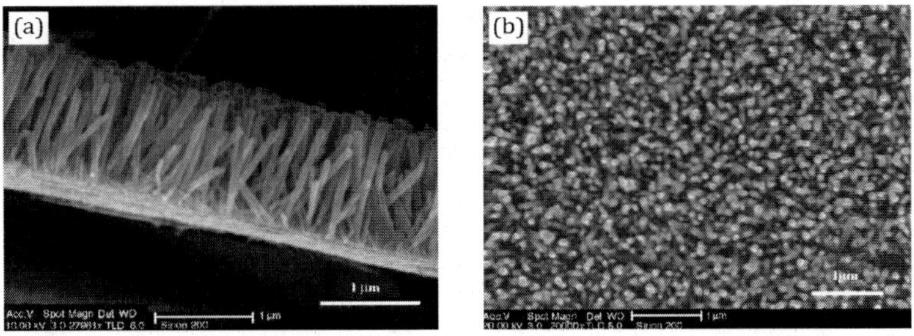

Figure 6.5 ZnO nanorod arrays on a silicon substrate using the MBE method. (a) Side view. (b) Top view. Reprinted from Wang, M. Y., et al. (2006), "Synthesis of well-aligned ZnO nanorod arrays with high opticalproperty via a low-temperature solution method," *J. Cryst. Growth*, 291, pp. 334–339, Copyright (2006), with permission from Elsevier.

6.1.6 Reactive Pulsed Laser Deposition (PLD) Method

In the reactive pulsed laser deposition (PLD) method [Sun et al., 2004; Bao et al., 2005; Fuge et al., 2009; Fusi et al., 2009], the pulsed laser produces gas composed of atoms, ions, and electrons from the target material. After a substrate is exposed to the emerging gas on one side, it is turned over. The slow formation of nano-elements then takes place on the back side of the film in the atmosphere. Figure 6.6 shows ZnO wires grown on a silicon substrate.

6.1.7 Geometrical Features

Figure 6.7 summarizes film classifications in terms of the geometry of the nano-elements, which in turn governs the mechanical behavior characteristics of the aggregated film.

The geometrical features are governed by three main factors:

1. Dimension
2. Direction
3. Shape (including lattice structure)

In addition, when we consider the aggregation of the elements, we look at

4. Arrangement
5. Global pattern
6. Number of layers

These factors control mechanical properties as well. Finally, the chemical composition of the elements an additionally important item.

These factors strongly affect not only the mechanical behavior of deformation and fracture, e.g., stiffness, anisotropy, yield stress, and fracture elongation (fracture strain), but also other physical behavior, e.g., thermal conductance.

In the following sections, we discuss characteristic mechanical behaviors, using as an example an aggregated film made by a dynamic oblique deposition (DOD) technique.

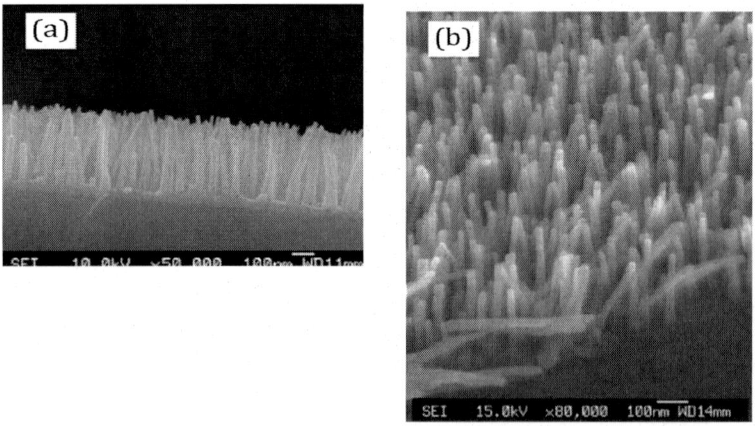

Figure 6.6 ZnO wires grown on a silicon substrate using the PLD method. (a) Side view. (b) Top view. Reprinted from Sun, Y., Fuge, G. M., and Ashfold, M. N. R. (2004) "Growth of aligned ZnO nanorod arrays by catalyst-free pulsed laser deposition methods," *Chem. Phys. Lett.*, 396, pp. 21–26, Copyright (2004), with permission from Elsevier.

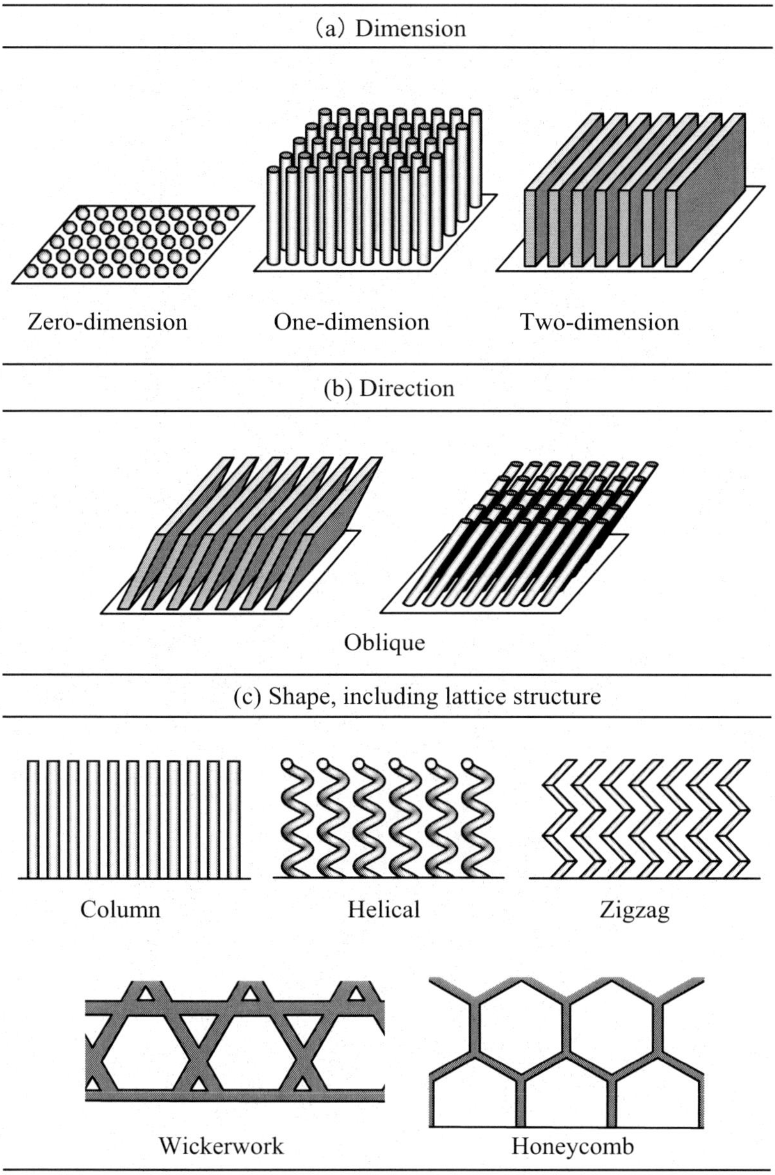
(a) Dimension
Zero-dimension
One-dimension
Two-dimension
(b) Direction
Oblique
(c) Shape, including lattice structure
Column
Helical
Zigzag
Wickerwork
Honeycomb

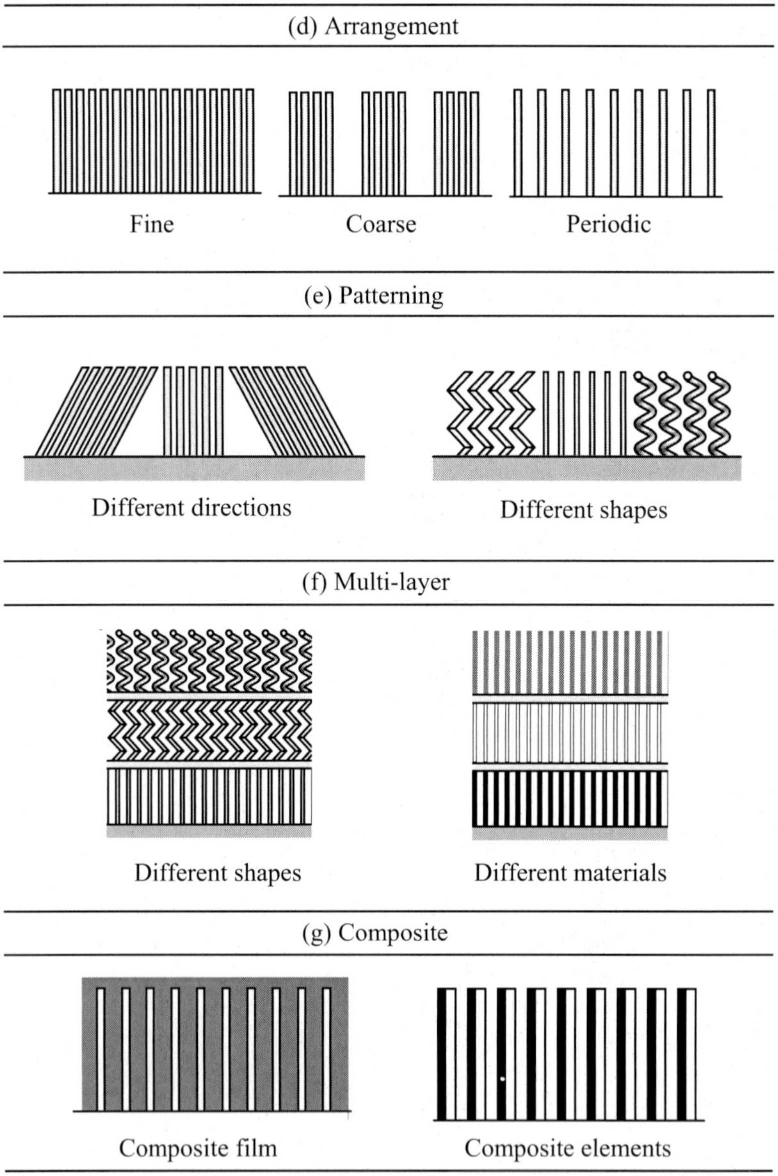

Figure 6.7 Classification of aggregated film in terms of geometry of nano-elements.

6.2 MECHANICAL PROPERTIES OF AGGREGATED FILM

6.2.1 Fabrication

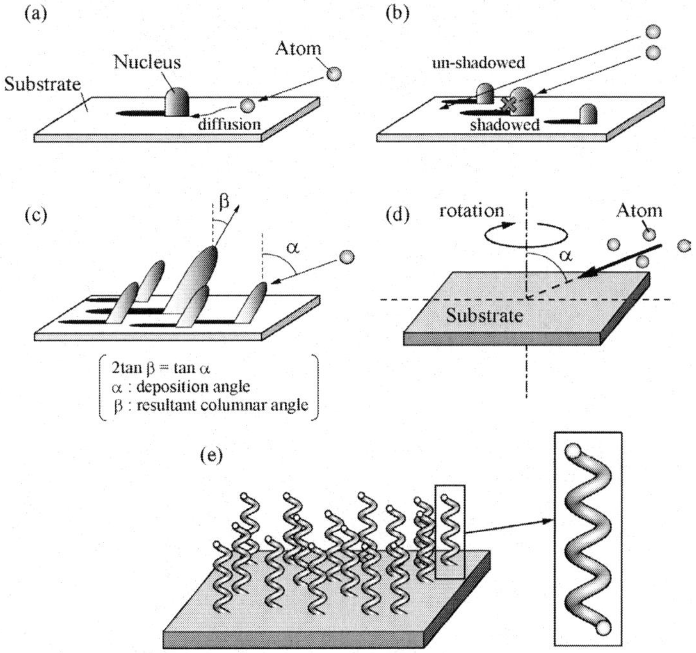

Figure 6.8 Schematic illustration of dynamic oblique deposition.

Figure 6.8 shows the process of forming an aggregated film using the DOD technique. As the atoms are deposited at a slant, numerous nano-meter islands nucleated on the substrate, as schematically illustrated in Fig. 6.8a. Once islands are formed, successive atoms cannot reach the other side of the oblique deposition, creating a shadow effect. Thus, each island grows in one direction and forms a slanted nano-bar, as shown in Fig. 6.8b,c. If the substrate is rotated during deposition, the growth direction of bar becomes twisted, and this results in a spring shape, as shown in Fig. 6.8d,e. By changing the rotation, we can readily control the shape and size of the twist. In other words, by coordinating the oblique angle and speed of deposition as well as the rotation of the substrate, we can create precisely sculptured nano-elements. A numerical simulation of this process has been developed [Suzuki and Taya, 2001b] to guide design shape. Figure 6.9 shows examples of helical, zigzag, columnar, and other shaped elements obtained through this technique.

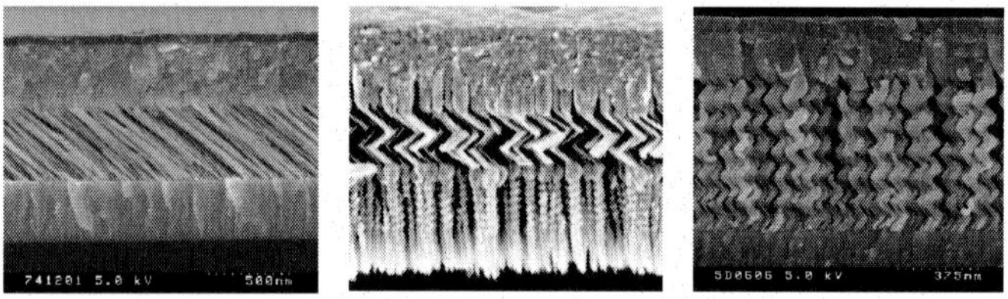

(a) (b) (c)

Figure 6.9 Thin films comprised of aggregated sculptured nano-elements. (a) Oblique columnar. (b) Zigzag. Reprinted with permission from Suzuki, M., and Taga, Y. (2001) "Integrated sculptured thin films," *Jpn. J. Appl. Phys.*, 40, pp. L358–L359. (c) Helical. Reprinted from Hirakata, H., et al. (2007) "Anisotropic deformation of thin films comprised of helical nanosprings," *Int. J. Solid Struct.*, 44, pp. 4030–4038, Copyright (2007), with permission from Elsevier.

6.2.2 Deformation Behavior of Aggregated Film and Single Helical

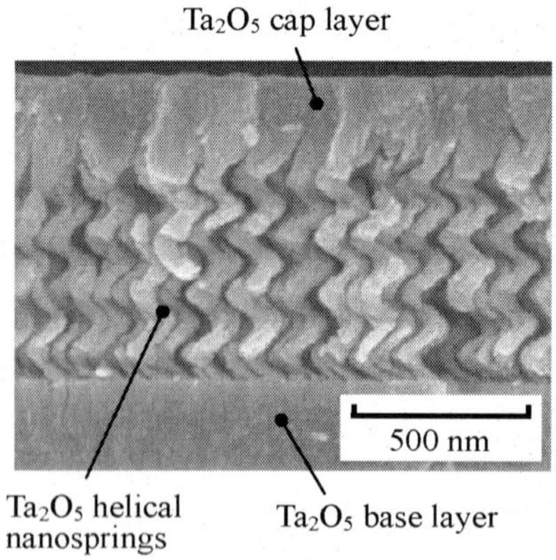

Figure 6.10 SEM micrograph of a Ta_2O_5 nanosprings thin film sandwiched between a Ta_2O_5 cap and a base layer. Reprinted from Sumigawa, T., et al. (2008) "Disappearance of stress singularity at interface edge due to nanostructured thin film," *Eng. Fract. Mech.*, 75, pp. 3073–3083, Copyright (2008), with permission from Elsevier.

Figure 6.10 shows a SEM micrograph of the vertical cross-section of a thin film comprised of tantalum pentoxide (Ta_2O_5) helical nanosprings. The nanosprings are made using an electron beam evaporator (deposition angle: 82°, vacuum: 8.0 × 10^{-3} Pa, deposition speed: 0.1 nm/s). Each element has a helical shape, where the height, radius, wire diameter, and rotation number are 560 nm, 90 ± 10 nm, 40 ± 10 nm, and 3.5, respectively. The density of elements (the number of springs per unit area) is 65 μm^{-2}.

Plural blocks (S = 1.9 – 13.2 μm^2, S: area of specimen) are carved out from the thin film by means of a focused ion beam (FIB), as shown in Fig. 6.11, where a cap layer of Ta_2O_5 is formed to apply the load.

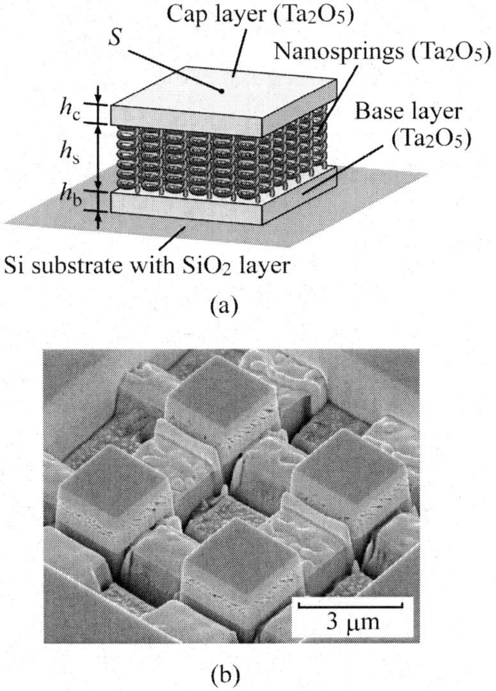

Figure 6.11 Configuration of specimen: a brick cut from the multi-layered thin films by FIB. (a) Schematic illustration of specimen. Reprinted from Sumigawa, T., et al. (2008) "Disappearance of stress singularity at interface edge due to nanostructured thin film," *Eng. Fract. Mech.*, 75, pp. 3073–3083, Copyright (2008), with permission from Elsevier. (b) SEM image.

The film is subjected to a vertical or lateral load by means of an atomic force microscope. The microscope can precisely control and measure the vertical and lateral forces, F_v and F_l, and the corresponding displacements of the diamond tip, δ_v and δ_l. Pushing a diamond tip into the cap layer, as illustrated in Fig. 6.12, allows us to examine the two-dimensional stiffness of the specimen [Sumigawa et al., 2008]. A conical tip with a large curvature (R = 1 μm) is used to avoid damaging the cap layer under vertical loading (Fig.6.12a), while a pyramid tip with a vertical angle of 60° is used to inhibit slipping at the point of contact under lateral loading (Fig.6.12b).

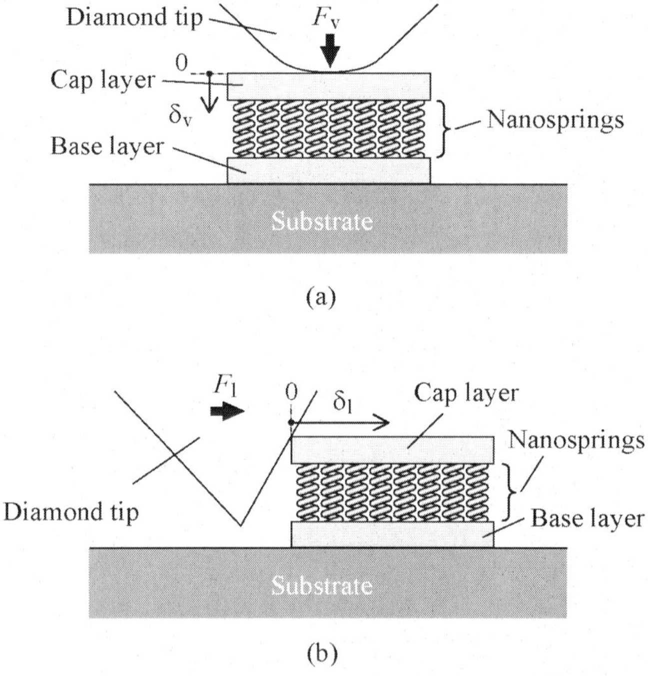

Figure 6.12 Loading methods for measuring the mechanical properties of the thin film composed of nanosprings. Reprinted from Sumigawa, T., et al. (2008) "Disappearance of stress singularity at interface edge due to nanostructured thin film," *Eng. Fract. Mech.*, 75, pp. 3073–3083, Copyright (2008), with permission from Elsevier. (a) Vertical loading method. (b) Lateral loading method.

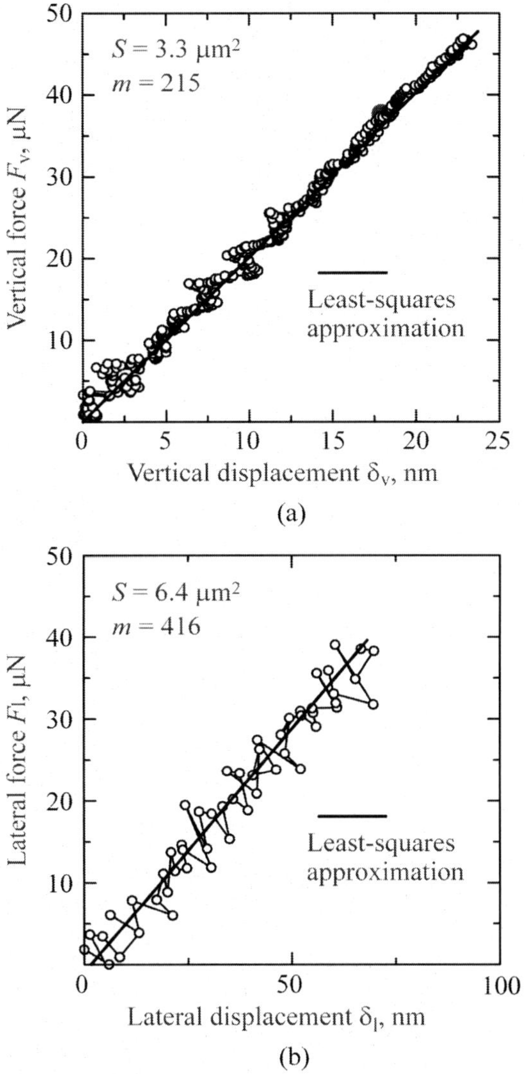

Figure 6.13 Load-displacement relationship. (a) Vertical loading test. (b) Lateral loading test. Reprinted from Sumigawa, T., et al. (2008) "Disappearance of stress singularity at interface edge due to nanostructured thin film," *Eng. Fract. Mech.*, 75, pp. 3073–3083, Copyright (2008), with permission from Elsevier.

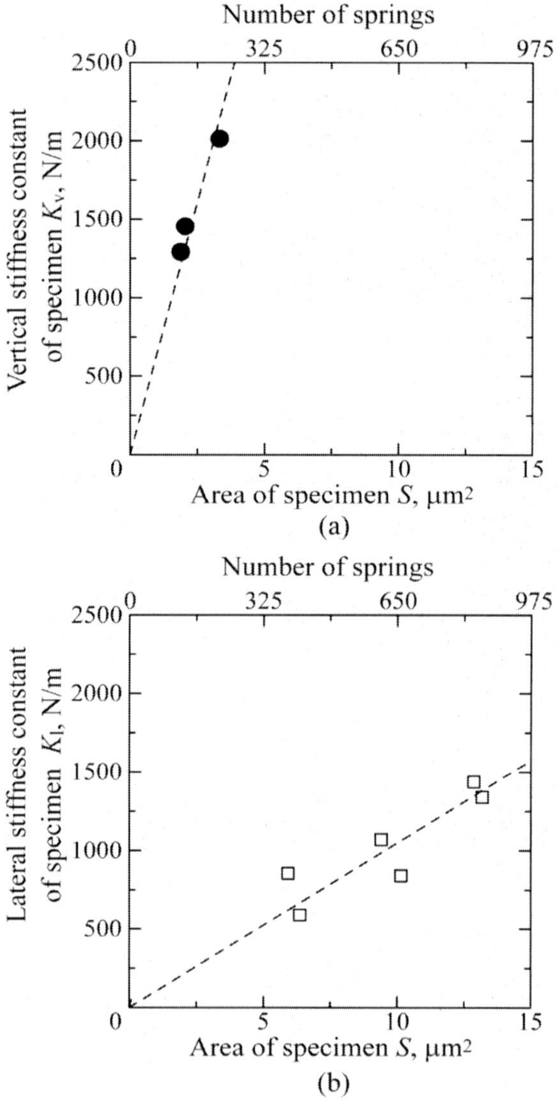

Figure 6.14 Relationship between the deformation stiffness constants, (a) K_v and (b) K_l, and the area of the specimen, S. Reprinted from Sumigawa, T., et al. (2008) "Disappearance of stress singularity at interface edge due to nanostructured thin film," *Eng. Fract. Mech.*, 75, pp. 3073–3083, Copyright (2008), with permission from Elsevier.

Figure 6.13a shows the relationship between the force, F_v, and the displacement, δ_v, in a specimen under vertical loading. Because the F_v–δ_v curve is linear and reversible, the vertical stiffness of the specimen, K_v (= F_v/δ_v), is evaluated by the slope. The stiffness is dependent on the specimen size (number of nanosprings), as shown in Figure 6.14a, which plots the results of several specimens with different area S. The stiffness of a single nanospring can be evaluated by

$$k_v = \frac{K_v}{m} \tag{6.1}$$

where m is the number of nanosprings (density times S) in the specimen, shown at the top of figure. Because the K_v–m relation has linearity, k_v is evaluated from the slope.

Similarly, the lateral stiffness of specimens, K_l, can be evaluated by the $F_l - \delta_l$ curve in Fig. 6.13b. That of a single nanospring k_l is derived from the slope in the $K_l - S$ ($K_l - m$) relation by

$$k_l = \frac{K_l}{m} \tag{6.2}$$

The magnitudes of k_v and k_l are listed in Table 6.1.

Table 6.1 Stiffness of thin films of Ta_2O_5 nanosprings. E': apparent Young's modulus; G': apparent shear modulus; E'/G': ratio of E' to G'; k_v: vertical stiffness of a single nanospring; k_l: lateral stiffness of a single nanospring.

	k_v, N/m	**k_l, N/m**	**E', GPa**	**G', GPa**	**E'/G'**
Ta_2O_5 nanosprings thin film	10.3	1.6	0.37	0.060	6.2
Ta_2O_5 solid thin film	–	–	117±10	47±4	2.5

In general, a helical spring has much less stiffness in the transverse direction than in the longitudinal direction. The ratio of k_v to k_l of a single helical spring is given approximately by the following equation [Wahl, 1963].

$$\frac{k_v}{k_l} = \frac{G}{E}\left\{1 + \frac{1}{12}\left(\frac{h_s}{r}\right)^2\left(1 + \frac{E}{2G}\right)\right\} \tag{6.3}$$

The strong deformational anisotropy and low stiffness of the element brings about characteristic mechanical properties of the film.

To help the readers understand, we introduce the apparent Young's modulus, E', and the apparent shear modulus, G', of a film by the following definition, although they are slightly different from the exact definition of the modulus, E and G, in continuum mechanics.

$$E' = K_V \frac{h_S}{S} \tag{6.4}$$

$$G' = K_1 \frac{h_S}{S} \tag{6.5}$$

where h_s is the height of the nanosprings. E' and G' of the nanosprings film are two to three orders lower than those of the Ta_2O_5 solid film (E = 117 GPa and G = 47 GPa) [Hirakata et al., 2007]. The ratio of E' to G' is 6.2, while that of the Ta_2O_5 solid film is about 2.5. The ratio of E to G of an isotropic solid film cannot go over 3.0 because it has the relation

$$E/G = 2(1+\nu) \tag{6.6}$$

where ν is the Poisson's ratio. This means that the nanosprings film has a strong anisotropy that is similar to its unidirectional reinforced composite.

6.3 DESIGN OF FILM PROPERTY ON THE BASIS OF ELEMENT CONFIGURATION

One of the major advantages of the DOD technique is that element configuration can be easily controlled by controlling the conditions of deposition and substrate motion. Figure 6.15 shows aggregated films comprised of different nanosprings [Hirakata et al., 2007], where Film B is the one discussed in the previous section. It is apparent that the springs in Film A are tighter and taller than those in Film B. The difference in helical configuration listed in the table in the figure brings about a change in the mechanical property of the spring, as shown in Table 6.1. The vertical stiffness is higher in Film B, while the lateral stiffness is lower than that of Film A.

This signifies that we can independently control vertical and lateral stiffness by tuning the helical configuration of the spring. In other words, we can design not only film stiffness (vertical and lateral) but also film anisotropy. Similarly, changing the shape of elements (e.g., from spring to zigzag bar), we can extend the design parameters.

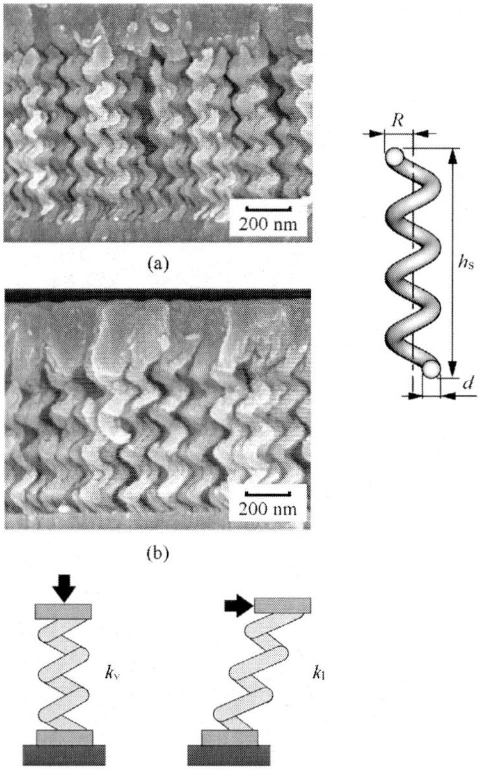

	n	h_s, nm	r, nm	d, nm	m/S, No./μm^2	k_l, N/m	k_v, N/m
Material A	5.5	690	50	40 ± 10	80	2.05 ± 0.05	6.89 ± 0.50
Material B	3.5	560	75	60 ± 10	65	1.66 ± 0.30	10.29 ± 0.72

Figure 6.15 SEM images of thin films consisting of Ta_2O_5 helical nanosprings with different configuration and their stiffness constants. (a) Film A. (b) Film B. Reprinted from Hirakata, H., et al. (2007) "Anisotropic deformation of thin films comprised of helical nanosprings," *Int. J. Solid Struct.*, 44, pp. 4030–4038, Copyright (2007), with permission from Elsevier.

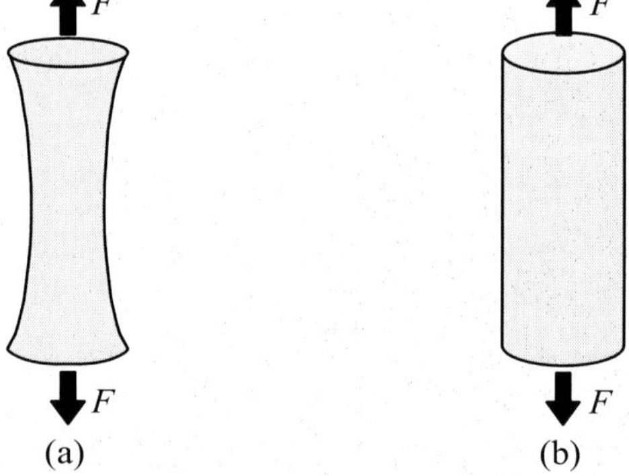

Figure 6.16 Schematic illustration of element shape. (a) Hourglass shape. (b) Columnar shape.

Other mechanical characteristics also depend on the shape of the nano-elements. For example, it is obvious that the fracture force is dependent on the stress distribution in the element. An element with an hourglass shape (Fig. 6.16a), with stress concentrated at the middle, can sustain a smaller load than one with a columnar shape (Fig. 6.16b) at the fracture. Thus, we can design the fracture strength of film by dealing with the stress concentration as well.

Moreover, it is easy to imagine that a spring tolerates larger extension than a straight column. This suggests that a film comprised of nanosprings can tolerate larger (apparent) elongation than a film comprised of nano-columns, because a spring converts vertical deformation into torsion. In fact, it has been reported that fracture strain is dependent on the configuration of the nano-elements on a film [Sumigawa et al., 2009]. The lower stiffness and higher fracture strain gives the film excellent reliability against thermal stress, which sometime causes fatal fracture in a micro-device. In a metal spring, this advantage might be enhanced by plastic deformation.

All of this indicates not only that it is possible to design the mechanical properties of an aggregated film but also that by controlling the element configuration we may be able to get specific mechanical behaviors that solid materials cannot attain.

Because the deformation properties and stress distribution of each element can be analyzed on the basis of conventional mechanics, FEM analysis of nano-elements is a strong tool in the design of film properties. Figure 6.17 shows the model of slant-bar element (E = 117 GPa and ν = 0.23), which has a columnar shape with the inclination angle β (= 0°, 15°, 30°, 45°, 60°, and 75°). The cap layer and substrate are treated as a rigid body for simplicity, and the forces in the x, y, and z directions, F_x, F_y, and F_z, are independently applied on the cap layer.

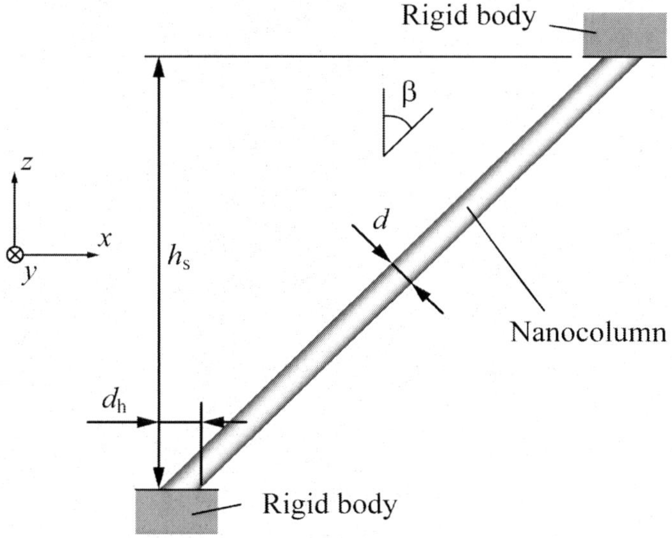

Figure 6.17 Analytical model in FEM analysis.

Figure 6.18 shows the apparent moduli of film, E'_{zz}, G'_{zx}, and G'_{zy}, converted from the element stiffness following Eqs. (6.4) and (6.5). The shaded area indicates the angle that we cannot fabricate with the DOD technique. The moduli decrease with the increase in the inclination angle because the elongated column element has a low bending rigidity. At β = 45°, E'_{zz} and G'_{zx} have same magnitude: E'/G' = 1. It should be noted that the ratio of E to G of an isotropic solid film is larger than 2 and cannot be 1 because of Eq. (6.6). Table 6.2 shows the experimental results (E'_{zz}, G'_{zx} and G'_{zy}) of aggregated film comprised of the Ta_2O_5 elements with β = 48°, which elucidates reasonable coincidence. This suggests that we can use numerical analysis based on continuum mechanics, such as FEM and the boundary element method (BEM) to design a film property through the nano-element configuration.

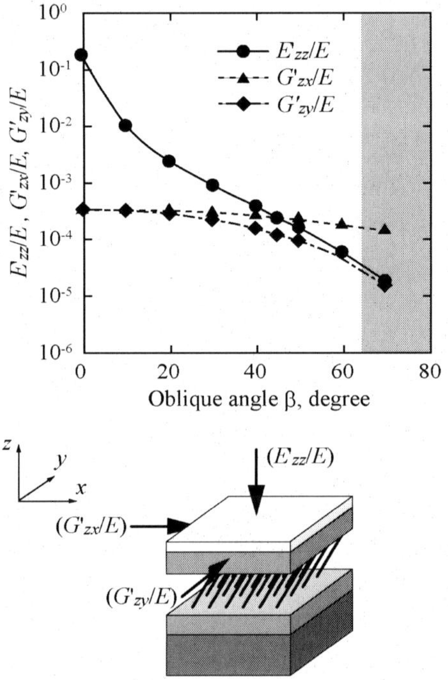

Figure 6.18 Relationships among the apparent moduli normalized by Young's modulus of solid film, E'_{zz}, G'_{zx}, and G'_{zy}, and the inclination angle, β.

Table 6.2 Apparent shear modulus and apparent Young's modulus of thin film comprised of Ta_2O_5 oblique nano-columns evaluated by FEM

	E'_{zz}, GPa	G'_{zx}, GPa	G'_{zy}, GPa
Thin film comprised of nano-columns	0.31 ± 0.07	0.33 ± 0.09	0.31 ± 0.06

6.4 DISAPPEARANCE OF STRESS CONCENTRATION/ SINGULARITY AT DISSIMILAR INTERFACE EDGE

As described in Chapters 2 and 5, the crack initiation at the interface edge and the crack growth along the interface are dominated by the stress concentration/singularity near the edge/tip. An approach based on fracture mechanics remains valid even at several tens of nanometer level. However, this approach might not be applicable to an aggregated thin film because the discreteness changes the stress condition. In this section, we examine the stress field of aggregated film sandwiched by dissimilar materials [Sumigawa et al., 2009].

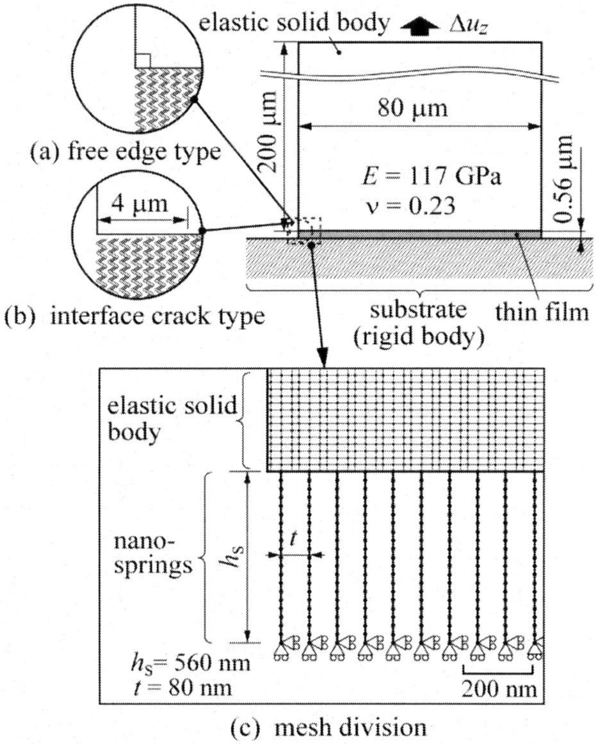

Figure 6.19 Analytical model in FEM analysis (spring model). Reprinted from Sumigawa, T., et al. (2008) "Disappearance of stress singularity at interface edge due to nanostructured thin film," *Eng. Fract. Mech.*, 75, pp. 3073–3083, Copyright (2008), with permission from Elsevier.

The stress distribution at the interface is calculated by the FEM using the model shown in Fig. 6.19, where the Ta_2O_5 nanosprings thin film (the stiffness is measured in Section 6.2) are represented by cylindrical beam elements inserted between a solid material (Ta_2O_5) and a substrate (rigid body). Two types of stress singularity are examined: One has a free edge where the surface-interface angle is

90° to 180°, as illustrated in Fig. 6.19a, and the other has an interface crack with a length of 4 μm where the surface-interface angle is 180° to 180° as illustrated in Fig. 6.19b. The conversion from the stiffness constants, k_l and k_v, to the beam is in the form

$$D_b = \sqrt{\frac{4k_l}{3k_v}} \cdot h_s \tag{6.7}$$

$$E_b = \frac{k_v^2}{k_l} \cdot \frac{3}{\pi h_s} \tag{6.8}$$

where h_s is the height of the nanospring. D_b is the beam diameter and E_b is the Young's modulus of beam, respectively.

Now we can look at a model without nanosprings for comparison. As shown in Fig. 6.20a, the interface edge (90°–180°) without the nanosprings clearly possesses singularity along the interface in the normal stresses, σ_x and σ_z, and the shear stress, τ_{zx}, normalized by the applied stress, σ_o. This signifies that the inserted aggregated film erases the stress singularity. On the other hand, the distributions in the edge with the nanosprings are nearly constant. Here, for comparison, the stress is defined as

$$\sigma = \frac{F_{el}}{A} \tag{6.9}$$

where F_{el} is the force acting on the element and A (t times unit thickness) is the area sustained by the element.

Figure 6.21 shows the distributions of the normalized stresses along the interface in a cracked model (180°–180°). In the direct mount (no aggregated film), the stresses severely concentrate near the crack tip, as shown in Fig. 6.21a. On the other hand, the aggregated film erases even the stress singularity of $r^{-1/2}$, as shown in Fig. 6.21b.

The stress concentration near the free edge of the interface is induced by the deformation mismatch of dissimilar materials. Since nanosprings are vertically oriented on the substrate and are separated from each other, the film has free in-plane deformation and in this case kills the mismatch.

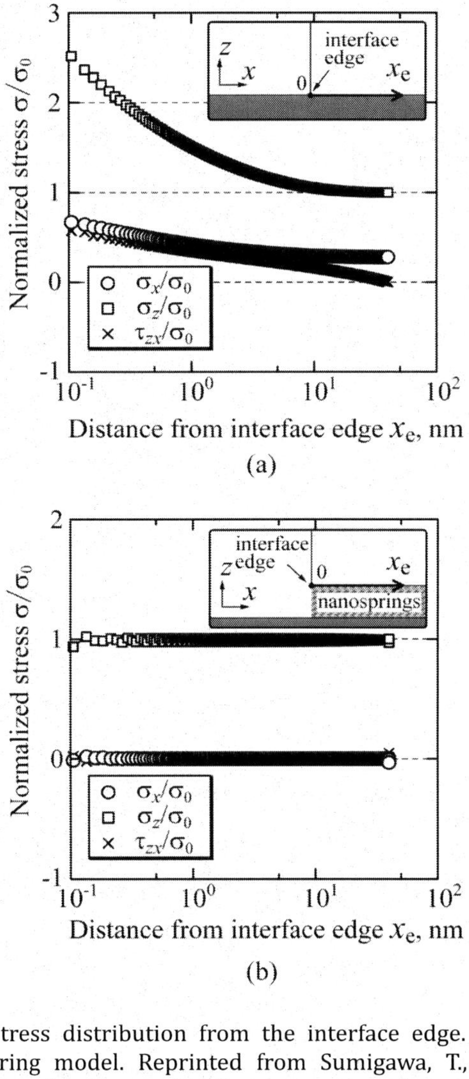

Figure 6.20 Stress distribution from the interface edge. (a) No-spring model. (b) Spring model. Reprinted from Sumigawa, T., et al. (2008) "Disappearance of stress singularity at interface edge due to nanostructured thin film," *Eng. Fract. Mech.*, 75, pp. 3073–3083, Copyright (2008), with permission from Elsevier.

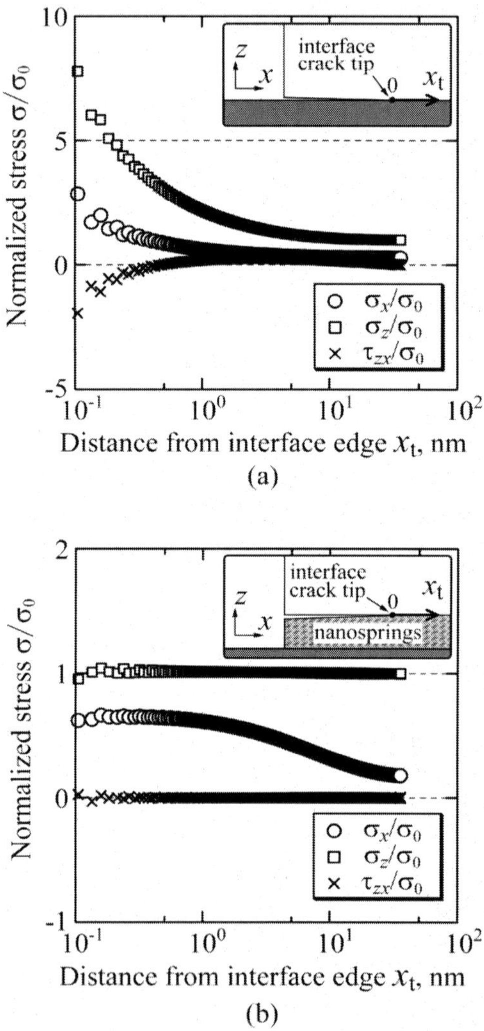

Figure 6.21 Stress distribution from the crack tip. (a) No-spring model. (b) Spring model. Reprinted from Sumigawa, T., et al. (2008) "Disappearance of stress singularity at interface edge due to nanostructured thin film," *Eng. Fract. Mech.*, 75, pp. 3073–3083, Copyright (2008), with permission from Elsevier.

6.5 CRACK INITIATION AND GROWTH ALONG A THIN FILM COMPRISED OF NANO-ELEMENTS

6.5.1 Crack Initiation

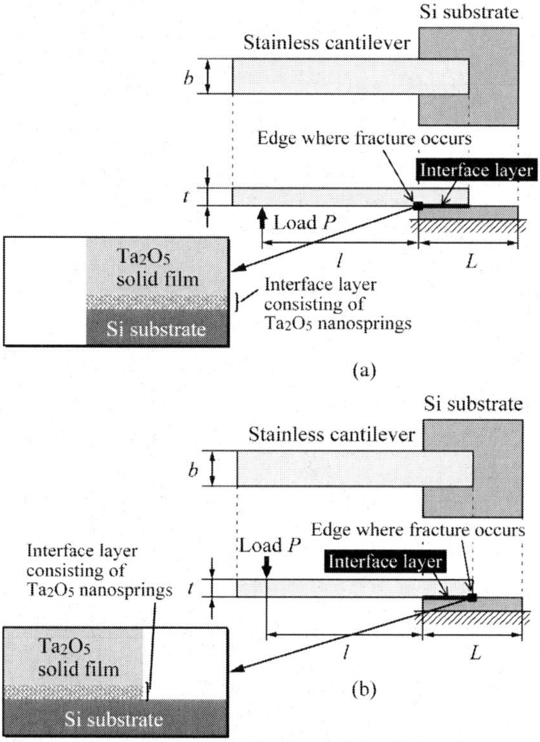

	L	b	t	l
A-1	2.27	1.58	1.92	17.5
A-2	2.17	2.09	1.87	11.0
A-3	1.37	1.77	1.90	15.5
B-1	1.74	1.68	1.90	15.5
B-2	1.85	1.95	1.89	16.0

Figure 6.22 Illustration of specimen configuration and loading method. Two types of tests possess different loading directions: (a) Type A test and (b) Type B test. Reprinted from Sumigawa, T., et al. (2009) "Effect of interface layer consisting of nanosprings on stress field near interface edge," *Eng. Fract. Mech.*, 76, pp. 1336–1344, Copyright (2009), with permission from Elsevier.

Figure 6.22 shows two types of specimens, Type A (Fig. 6.22a) and Type B (Fig. 6.22b). Although the specimen shapes are the same, the loading directions are not. The load is applied to the lower (or upper) face of the stainless steel cantilever at room temperature in an air (40 ± 15% R.H.) by means of a loading machine with an electromagnetic actuator, and a crack initiates at the left (or right) edge of the interface layer [Kitamura et al., 2003; Kitamura et al., 2002]. Figure 6.23 schematically illustrates the stress distributions near the crack initiation site (interface edge) in the specimen. No stress concentration is expected.

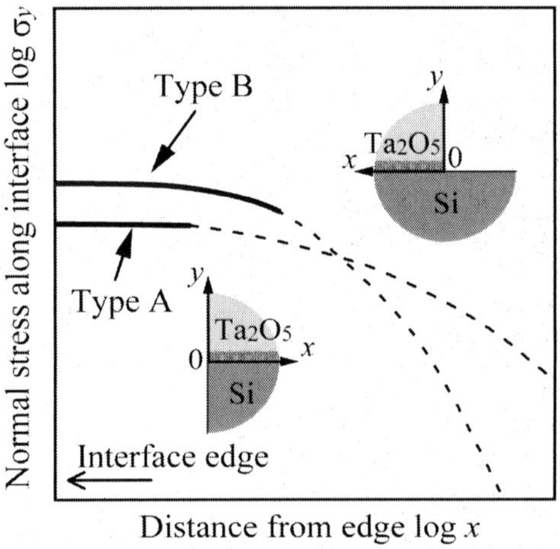

Figure 6.23 Schematic illustration of stress distributions near the interface edges in Types A and B. A Ta_2O_5 solid layer is mounted to the Si substrate with an interface layer consisting of nanosprings. Reprinted from Sumigawa, T., et al. (2009) "Effect of interface layer consisting of nanosprings on stress field near interface edge," *Eng. Fract. Mech.*, 76, pp. 1336–1344, Copyright (2009), with permission from Elsevier.

Figure 6.24 shows typical load-displacement curves (P – u_y relationships) in Types A (specimen A-1) and B (specimen B-1), respectively. The displacement at the loading point, u_y, increases linearly as the applied load, P, increases. The brittle fracture behavior indicates that the interface crack is initiated at the maximum load where the load sharply drops. Figure 6.25 shows the fracture surface near the edge in a Type A specimen, which signifies that the crack initiates within the nanosprings thin film.

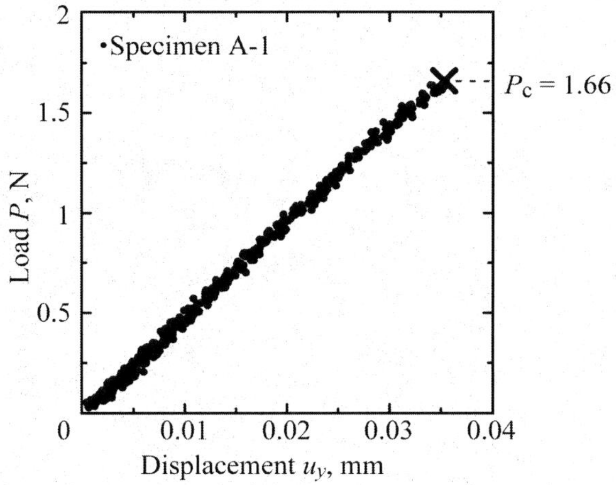

(a)

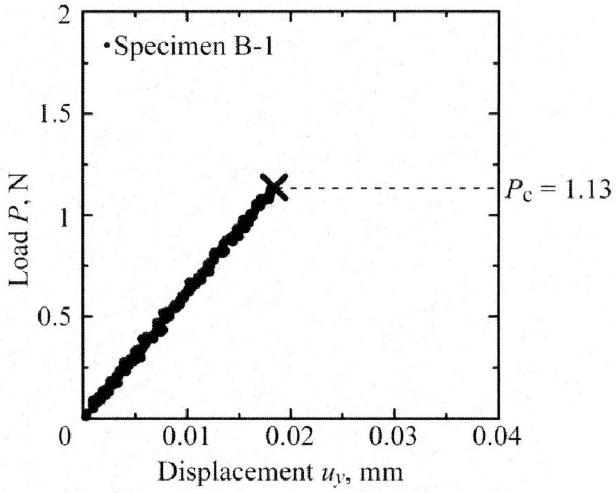

(b)

Figure 6.24 Load-displacement curves in two types of tests on the specimens with the nanosprings thin film: (a) Type A and (b) Type B. P_c indicates the critical load, at which an interface crack is initiated. Reprinted from Sumigawa, T., et al. (2009) "Effect of interface layer consisting of nanosprings on stress field near interface edge," *Eng. Fract. Mech.*, 76, pp. 1336–1344, Copyright (2009), with permission from Elsevier.

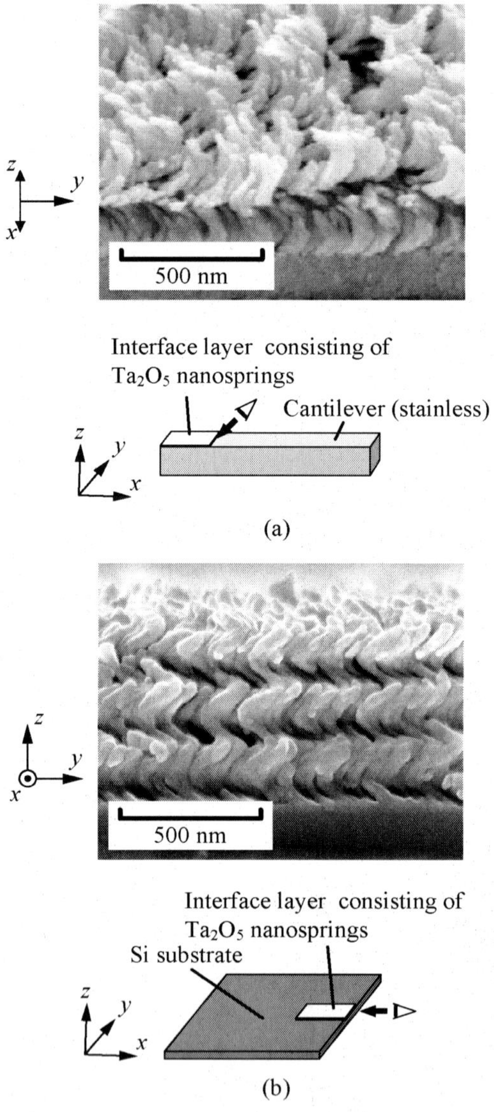

Figure 6.25 SEM observation of fracture surface. (a) Upper fracture surface. (b) Lower fracture surface. Reprinted from Sumigawa, T., et al. (2009) "Effect of interface layer consisting of nanosprings on stress field near interface edge," *Eng. Fract. Mech.*, 76, pp. 1336–1344, Copyright (2009), with permission from Elsevier.

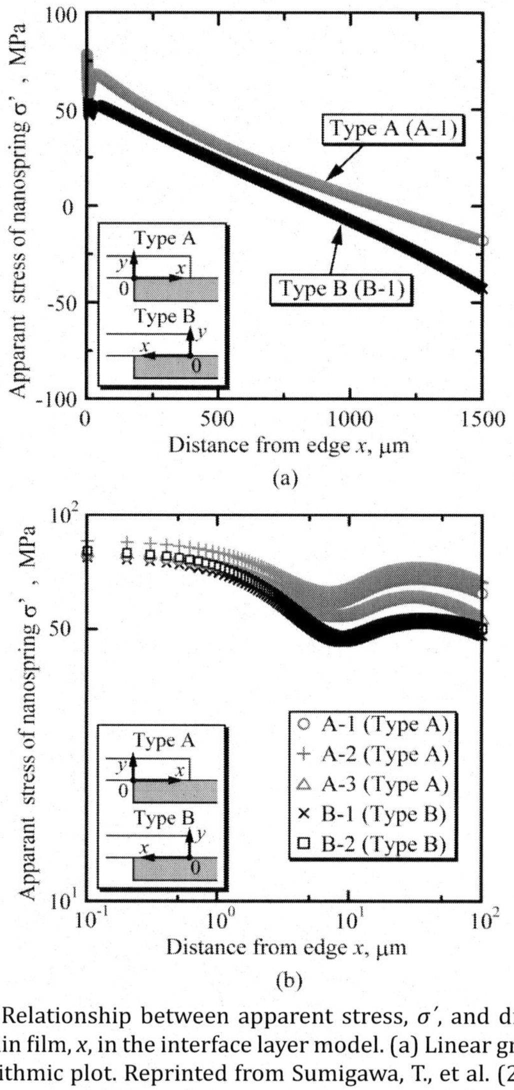

Figure 6.26 Relationship between apparent stress, σ', and distance from the edge of thin film, x, in the interface layer model. (a) Linear graph plot. (b) Double logarithmic plot. Reprinted from Sumigawa, T., et al. (2009) "Effect of interface layer consisting of nanosprings on stress field near interface edge," *Eng. Fract. Mech.*, 76, pp. 1336–1344, Copyright (2009), with permission from Elsevier.

Figure 6.26 shows the relationship calculated by means of the FEM between the stress, σ', and the distance from the edge of the nanosprings thin film, x, at the crack initiation in the model with nanospring thin film. The stress distributions away from the edge differ between the two test types, as shown in Fig. 6.26a, while the stress distribution curves of all specimens show a good agreement near the interface edge, as shown in the magnified view near the edge in Fig. 6.26b. This signifies that the stress at the edge dominates the crack initiation.

6.5.2 Crack Growth

A four-point bend specimen with a different crack length is made of (glass plate) / (Ta_2O_5 solid film) / (aggregated film comprised of Ta_2O_5 nanosprings) / (silicon substrate), as shown in Fig. 6.27a. The crack length can be monitored *in situ* during the experiment (at room temperature (298 K) in an air (40 ± 15% R.H.)) through the glass by means of an optical microscope, as shown in Fig. 6.27b, because the upper Ta_2O_5 layer is thin enough.

The crack begins to grow in the nanospring film without any nonlinear deformation in the load-displacement curve when the load reaches the critical value ($P = P_C$), namely brittle cracking.

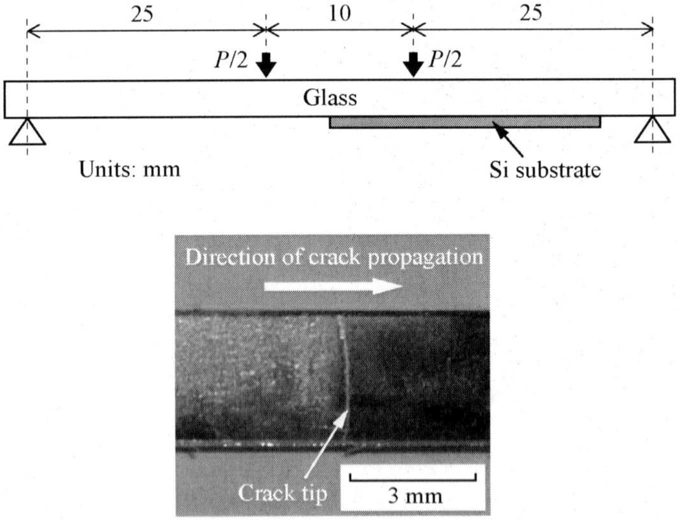

Figure 6.27 Specimen configuration and test style (four-point bending test).

Figure 6.28 shows the stress distribution near the crack tip at the critical load (specimens C-1, C-2, and C-3). There is no strong stress singularity in all cases and only a slight difference in the stress at crack tip among the three specimens. This indicates that the crack growth is governed by the stress at the crack tip regardless of the crack length. The stress intensity factor is not the governing fracture parameter anymore. Moreover, the critical stresses correspond with the one for the crack initiation superimposed. This shows that there is no difference in the critical stresses at the crack tip (crack growth) and the interface edge (crack initiation). In short, a unified critical apparent stress, σ'_c (σ'_c = 79 ± 6 MPa), causes the fracture of the first nanospring at the edge/crack tip and brings about the crack initiation/growth. This fact proves the disappearance of stress singularity at the crack tip as well.

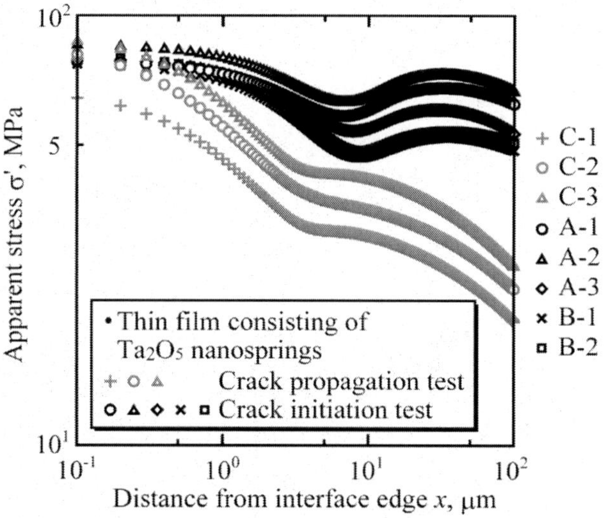

Figure 6.28 Stress conditions near the crack tip and near the interface edge.

References

Abdi, Y., Mohajerzadeh, S., Hoseinzadegan, H., and Koohsorkhi, J. (2006) Carbon nanostructures on silicon substrates suitable for nanolithography, *Appl. Phys. Lett.*, 88, Article number 053124.

Bao, Q., Chen, C., Wang, D., Ji, Q., and Lei, T. (2005) Pulsed laser deposition and its current research status in preparing hydroxyapatite thin films, *Appl. Surf. Sci.*, 252, pp. 1538–1544.

Bougrioua, Z., Gibart, P., Calleja, E., Jahn, U., Trampert, A., Ristic, J., Utrera, M., and Nataf G. (2007) Growth of freestanding GaN using pillar-epitaxial lateral overgrowth from GaN nanocolumns, *J. Cryst. Growth*, 309, pp. 113–120.

Cao, X., Lan, X., Zhao, C., Shen, W., Yao, D., and Gao, W. (2007) General wet route for the growth of regular anisotropic, *J. Cryst. Growth*, 306, pp. 225–232.

Chu, S. Z., Wada, K., Inoue, S., and Todoroki, S. (2003) Fabrication and characteristics of nanostructures on glass by Al anodization and electrodeposition, *Electrochim. Acta*, 48, pp. 3147–3153.

Chung, Y.-W., Leu, I.-C., Lee, J.-H., Yen, J.-H., and Hon, M.-H. (2007) Fabrication of egg-shell-roofed macroporous nickel films by a template-mediated electrodeposition process, *Electrochim. Acta*, 53, pp. 1703–1707.

Dyke, L. S. V., and Martin, C. R. (1990) Electrochemical investigations of electronically conductive polymers. 4. Controlling the supermolecular structure allows charge transport rates to be enhanced, *Langmuir*, 6, pp. 1118–1123.

Ekanayake, E. M. I. M., Preethichandra, D. M. G., and Kaneto, K. (2007) Polypyrrole nanotube array sensor for enhanced adsorption of glucose oxidase in glucose biosensors, *Biosen. Bioelectron.*, 23, pp. 107–113.

Elias, J., Tena-Zaera, R., and Levy-Clement, C. (2008) Electrochemical deposition of ZnO nanowire arrays with tailored dimensions, *J. Electroanal. Chem.*, 621, pp. 171–177.

Fuge, G. M., Holmes, T. M. S., and Ashfold, M. N. R. (2009) Ultrathin aligned ZnO nanorod arrays grown by a novel diffusive pulsed laser deposition method, *Chem. Phys. Lett.*, 479, pp. 125–127.

Fusi, M., Russo, V., Casari, C. S., Bassi, A. L., and Bottani, C. E. (2009) Titanium oxide nanostructured films by reactive pulsed laser deposition, *Appl. Surf. Sci.*, 255, pp. 5334–5337.

Guo, B., Liu, Z., Hong, L., and Jiang, H. (2005) Sol gel derived photocatalytic porous TiO_2 thin films, *Surf. Coat. Technol.*, 198, pp. 24–29.

Hersee, S. D., Sun, X., and Wang, X. (2006) The controlled growth of GaN nanowires, *Nano Lett.*, 6(8), pp. 1808–1811.

Hirakata, H., Matsumoto, S., Takemura, M., Suzuki, M., and Kitamura, T. (2007)

Anisotropic deformation of thin films comprised of helical nanosprings, *Int. J. Solids Struc.*, 44, pp. 4030–4038.

Hong, S.-H., Han, K.-S., Lee, H., Cho, J. U., and Kim, Y. K. (2007) Fabrication of nano-porous structure on silicon substrate using nanoimprint lithography with an anodic aluminum oxide nano-template, *Jpn. J. Appl. Phys.*, 46(9B), pp. 6375–6377.

Hu, X., Masuda, Y., Ohji, T., and Kato, K. (2009) Low-temperature fabrication of porous and transparent ZnO films with hybrid structure by self-hydrolysis method, *Thin Solid Films*, 518, pp. 638–641.

Jin, Z., Zhou, H.-J., Jin, Z.-L., Savinell, R. F., and Liu, C.-C. (1998) Application of nano-crystalline porous tin oxide thin film for CO sensing, *Sens. Actuators B*, 52, pp. 188–194.

Kato, T., Jeong, G.-H., Hirata, T., and Hatakeyama, R. (2004) Structure control of carbon nanotubes using radio-frequency plasma enhanced chemical vapor deposition, *Thin Solid Films*, 457, pp. 2–6.

Kim, Y. H., Lee, J. Y., Lee, S.-H., Oh, J.-E., and Lee, H. S. (2005) Synthesis of aligned GaN nanorods on Si (111) by molecular beam epitaxy, *Appl. Phys.*, A80, pp. 1635–1639.

Kipshidze, G., Yavich, B., Chandolu, A., Yun, J., Kuryatkov, V., Ahmad, I., Aurongzeb, D., Holtz, M., and Temkin, H. (2005) Controlled growth of GaN nanowires by pulsed metalorganic chemical vapor deposition, *Appl. Phys. Lett.*, 86, Article number 033104.

Kitamura, T., Hirakata H., and Itsuji, T. (2003) Effect of residual stress on delamination from interface edge between nano-films, *Eng. Fract. Mech.*, 70, pp. 2089–2101.

Kitamura, T., Shibutani, T., and Ueno, T. (2002) Crack initiation at free edge of interface, *Eng. Fract. Mech.*, 69, pp.1289–1299.

Kohl, A. T., Mimna, R., Shick, R., Rhodes, L., Wang, Z. L., and Kohl, P. A. (1999) Low k, porous methyl silsesquioxane and spin-on-glass, *Electrochem. Solid-State Lett.*, 2(2), pp. 77–79.

Lai, Y., Lin, C., Wang, H., Huang, J., Zhuang, H., and Sun, L. (2008) Superhydrophilic–superhydrophobic micropattern on TiO_2 nanotube films by photocatalytic lithography, *Electrochem. Comm.*, 10, pp. 387–391.

Lee, P.-S., Lee, O.-J., Hwang, S.-K., Jung, S.-H., Jee, S. E., and Lee, K.-H. (2005) Vertically aligned nanopillar arrays with hard skins using anodic aluminum oxide for nano imprint lithography, *Chem. Mater.*, 17, pp. 6181–6185.

Lee, W., Jeong, M.-C., Kim, M. J., and Myoung, J.-M. (2007) Field emission characteristics of ZnO nanoneedle array cell under ultraviolet irradiation, *Phys. Lett. A*, 370, pp. 345–350.

Malac, M., and Egerton, R. F. (2001) Observation of the microscopic growth mechanism of pillars and helices formed by glancing-angle thin film deposition, *J. Vac. Sci. Technol.*, A19(1), pp. 158–166.

Melechko, A. V., Merkulov, V. I., McKnight, T. E., Guillorn, M. A., Klein, K. L., Lowndes, D. H., and Simpson, M. L. (2005) Vertically aligned carbon nanofibers and related structures: controlled synthesis and directed assembly, *J. Appl. Phys.*, 97, Article number 041301.

Meyyappan, M., Delzeit, L., Cassell, A., and Hash, D. (2003) Carbon nanotube growth by PECVD: a review, *Plasma Sources Sci. Technol.*, 12, pp. 205–216.

Mohan, P., Motohisa, J., and Fukui, T. (2005) Controlled growth of highly uniform, axial/radial direction-defined, individually addressable InP nanowire arrays, *Nanotechnol.*, 16, pp. 2903–2907.

Qu, L., Du, F., and Dai, L. (2008) Preferential syntheses of semiconducting vertically aligned single-walled carbon nanotubes for direct use in FETs, *Nano Lett.*, 8(9), pp. 2682–2687.

Robbie, K., Brett, M. J., and Lakhtakia, A. (1995) First thin film realization of a helicoidal bianisotropic medium, *J. Vac. Sci. Technol.*, A13(6), pp. 2991–2993.

Robbie, K., and Brett, M. J. (1997) Sculptured thin films and glancing angle deposition: growth mechanics and applications, *J. Vac. Sci. Technol.*, A15(3), pp. 1460–1465.

Sanchez-Paramo, J., Calleja, J. M., Sanchez-Garca, M. A., Calleja, E., and Jahn, U. (2002) Structural and optical characterization of intrinsic GaN nanocolumns, *Physica E*, 13, pp. 1070–1073.

Shin, S. W., Lee, S. G., Lee, J., Whang, C. N., Lee, J. H., Choi, I.-H., Kim, T. G., and Song, J. H. (2005) Ion-beam nano-patterning by using porous anodic alumina as a mask, *Nanotechnol.*, 16, pp. 1392–1395.

Sumigawa T., Hirakata, H., Takemura, M., Matsumoto, S., Suzuki, M., and Kitamura, T. (2008) Disappearance of stress singularity at interface edge due to nanostructured thin film, *Eng. Fract. Mech.*, 75, pp. 3073–3083.

Sumigawa, T., Sueda, T., Futamura, Y., Suzuki, M., and Kitamura, T. (2009) Effect of interface layer consisting of nanosprings on stress field near interface edge, *Eng. Fract. Mech.*, 76, pp. 1336–1344.

Sun, Y., Fuge, G. M., and Ashfold, M. N. R. (2004) Growth of aligned ZnO nanorod arrays by catalyst-free pulsed laser deposition methods, *Chem. Phys. Lett.*, 396, pp. 21–26.

Suzuki, M., and Taga, Y. (2001a) Integrated sculptured thin films, *Jpn. J. Appl. Phys.*, 40, pp. L358–L359.

Suzuki, M., and Taga, Y. (2001b) Numerical study of the effective surface area of obliquely deposited thin films, *J. Appl. Phys.*, 90, pp. 5599–5605.

Tomioka, K., Mohan, P., Noborisaka, J., Hara, S., Motohisa J., and Fukui, T. (2007) Growth of highly uniform InAs nanowire arrays by selective-area MOVPE, *J. Cryst. Growth*, 298, pp. 644–647.

Vazquez, M., Hernandez-Velez, M., Asenjo, A., Navas, D., Pirota, K., Prida, V., Sanchez, D., and Baldonedo, J. L. (2006) Preparation and properties of novel magnetic composite nanostructures: arrays of nanowires in porous membranes, *Physica B*, 384, pp. 36–40.

Wahl, A. M. (1963) *Mechanical Springs*, 2nd ed., McGraw-Hill, New York.

Wang, M., Ye, C.-H., Zhang, Y., Hua G.-M, Wang, H.-X., Kong, M.-G., and Zhang, L.-D. (2006) Synthesis of well-aligned ZnO nanorod arrays with high optical property via a low-temperature solution method, *J. Cryst. Growth*, 291, pp. 334–339.

Yao, J.-L., Tang, J., Wu, D.-Y., Sun, D.-M., Xue, K.-H., Ren, B., Mao, B.-W., and Tian, Z.-Q. (2002) Surface enhanced Raman scattering from transition metal nano-wire array and the theoretical consideration, *Surf. Sci.*, 514, pp. 108–116.

Zang, K. Y., Wang, Y. D., Chua, S. J., and Wang, L. S. (2005) Nanoscale lateral epitaxial overgrowth of GaN on Si (111), *Appl. Phys. Lett.*, 87, Article number 193106.

Zhu, Y. C., and Ding, C. X. (1999) Plasma spraying of porous nanostructured TiO_2 film, *Nanostruct. Mater.*, 11(3), pp. 319–323.

Chapter 7

STRENGTH IN ATOMIC COMPONENTS

7.1 ATOMIC MECHANICS

Recent remarkable progress in manufacturing technology enables us to manufacture atomic-scale materials, the structure of which we can precisely control at an atom-by-atom level, e.g., carbon nanotubes (CNTs), gold nanowires, and atomic chains consisting of gold atoms arranged in a line. In these atomic-scale materials, even a slight displacement of atoms can lead to critical malfunction or failure of a component. This is especially true of surfaces and interfaces, where disarrangement of the atoms can affect the mechanical behavior of the materials as well as other of their physical properties. This suggests that conventional concepts of material strength based on continuum mechanics cannot be applied to atomic components. Hence, it is essential to understand the intrinsic strength of these materials from atomistic and electronic points of view. However, it is, in general, quite difficult to carry out precise experiments on an atomic scale.

Recent dramatic progress in computing power enables us to analyze the mechanical behavior of materials at the atomic level by using numerical calculations such as molecular dynamics (MD) simulations. Since we can freely impose arbitrary atomic structures, arbitrary loading and constraint, and arbitrary boundary conditions to the target nano-components, detailed mechanical behavior

Fracture Nanomechanics
Edited by Takayuki Kitamura, Hiroyuki Hirakata, Takashi Sumigawa and Takahiro Shimada

www.panstanford.com

of atomic components can be investigated thoroughly in this computational experiment.

In this chapter, we briefly review a few special experimental techniques for atomic components. Then we describe the fundamental concept and methodology of molecular dynamics simulations.

7.2 EXPERIMENTAL MECHANICAL TESTING FOR ATOMIC COMPONENTS

7.2.1 Mechanical Testing of a Gold Atomic Chain Using a Scanning Electron Microscope (STM)

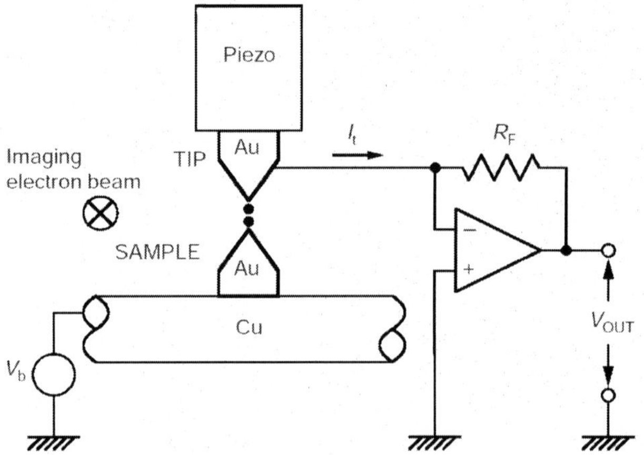

Figure 7.1 Loading device using a scanning tunneling microscope (STM) with a specimen stage built in an ultra-high vacuum (UHV) electron microscope. Reprinted by permission from Macmillan Publishers Ltd: Ohnishi, H., et al. (1998) "Quantized conductance through individual rows of suspended gold atoms," Nature, 395, pp. 780–783, copyright 1998.

Figure 7.1 shows a loading device that uses a scanning tunneling microscope (STM) with a specimen stage built in an ultrahigh-vacuum (UHV) high-resolution transmission electron microscope (HRTEM) [Ohnishi et al., 1998]. The UHV HRTEM (at 10^{-8} Pa) makes it possible the clean fabrication of gold tips and samples *in situ*. A mechanically sharpened gold tip is positioned close to a gold island that has been deposited on a thin copper wire (see Fig. 7.1). This gold tip can be displaced in three dimensional directions using a shear-type piezoelectric transducer. The gold tip is dipped into the gold island and slowly withdrawn. Structural change in gold atoms can be observed *in situ* during tension and the conductance can be measured simultaneously.

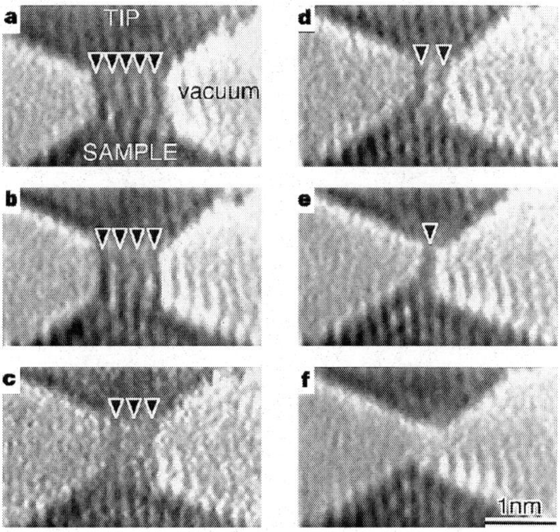

(a)

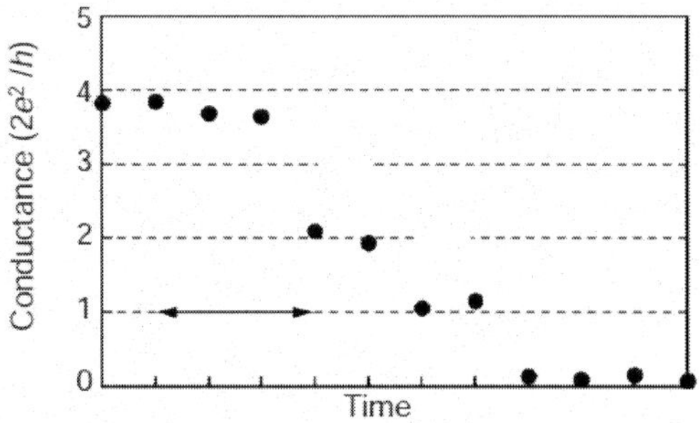

(b)

Figure 7.2 (a) Electron microscope images of a gold bridge formed between the gold tip (top) and gold substrate (bottom) thinned from **a** to **e** and ruptured at **f**. (b) Quantized conductance of a single and a double strand of gold atoms . Reprinted by permission from Macmillan Publishers Ltd: Ohnishi, H., et al. (1998) "Quantized conductance through individual rows of suspended gold atoms," Nature, 395, pp. 780–783, copyright 1998.

Figure 7.2a shows the electron microscope images of a gold bridge formed between an Au wire and tip during the tension. The bridge thins in Figs. 7.2a **a–d** and forms an atomic chain at **e**. Finally, it breaks in Fig. 7.2a **f**. Figure 7.2b shows the change in conductance of an atomic chain while this is happening. The atomic chain clearly exhibits quantized conductance.

7.2.2 Mechanical testing of carbon nanotubes (CNTs) using atomic force microscope (AFM)

Figure 7.3 shows a single-wall carbon nano-tube (SWCNT) mounted between atomic force microscope (AFM) tip and substrate [Yu et al. 2000]. The tensile load is applied to the top cantilever by a connected pico-motor and the magnitude of load is measured by the bending displacement. The deflection of the cantilever (force) and the length change of the CNT are simultaneously obtained by the scanning electron microscope (SEM).

Figures 7.4a,b show the SEM micrograph and the stress-strain curve, respectively. From the images, the fracture clearly takes place in the middle of the CNT. These point out that the CNT possesses extremely high Young's modulus of 950 GPa and tensile strength of 11–63 GPa.

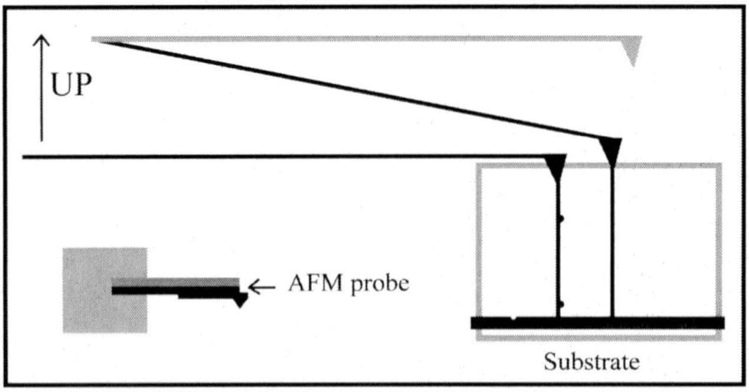

Figure 7.3 A schematic showing an overview of the tensile-loading experiment. In the schematic, the gray cantilever indicates where the cantilever would be if no rope were attached on the AFM tip. Redrawn from Yu, M. F., et al. (2000) "Strength and breaking mechanism of multiwalled carbon nanotubes under tensile load," *Science*, 287, pp. 637–640.

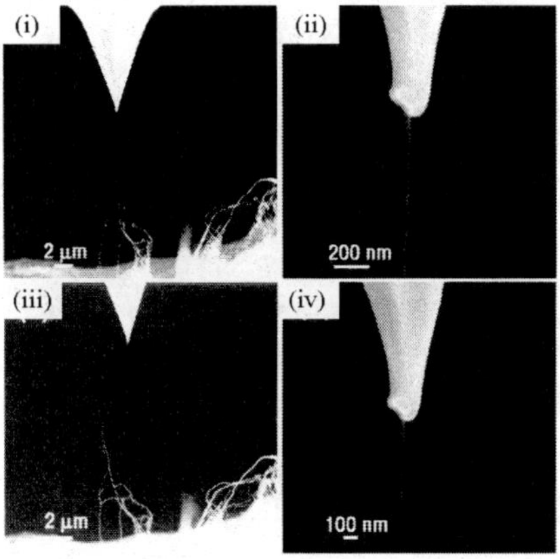

(a)

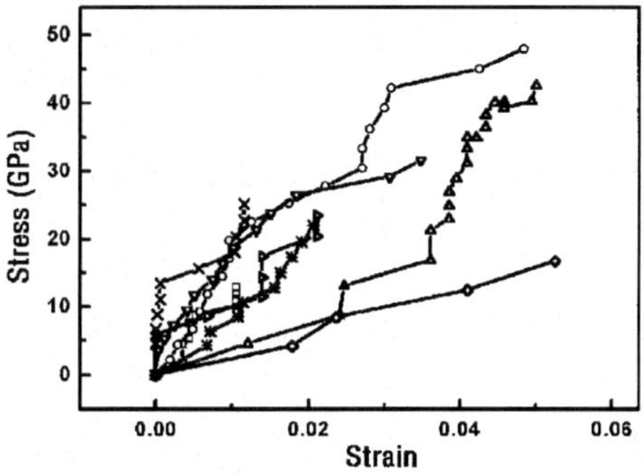

(b)

Figure 7.4 (a) Fracture of a single wall carbon nanotue (CNT) captured in SEM images. (b) Stress-strain curves of CNTs. From Yu, M. F., et al. (2000) "Strength and breaking mechanism of multiwalled carbon nanotubes under tensile load," *Science*, 287, pp. 637–640. Reprinted with permission from AAAS.

7.2.3 *In Situ* Indentation Test of Nano-particles Using a Transmission Electron Microscope (TEM)

The compression behavior of nano-particles is examined by means of an indentation device furnished in a transmission electron microscope (TEM). Figure 7.5 shows a schematic illustration of the *in situ* indentation holder. The construction of the experimental equipment is almost the same as that described in Section 5.2.3.

Figure 7.6a–c shows the detailed fracture process of a CdS nanoparticle [Shan et al., 2008]. The stiffness and strength of the nanoparticle are successfully evaluated (see Fig. 7.6f).

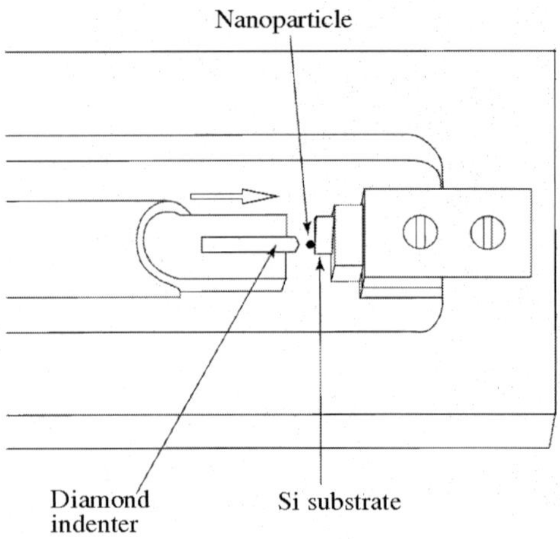

Figure 7.5 Schematic illustration of an in situ indentation holder in a transmission electron microscope (TEM). The electron beam is perpendicular to the plane formed by the particle and the diamond tip.

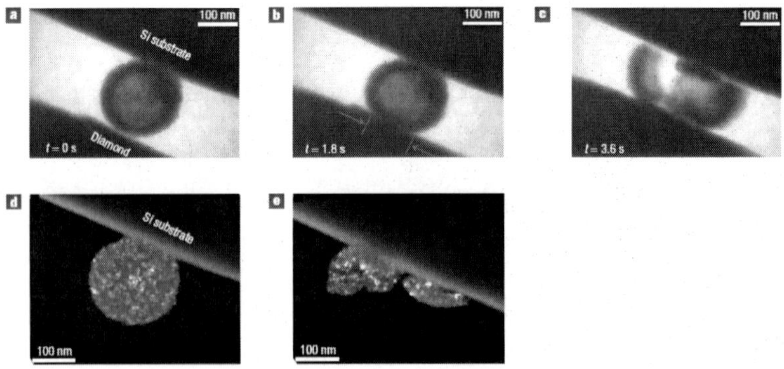

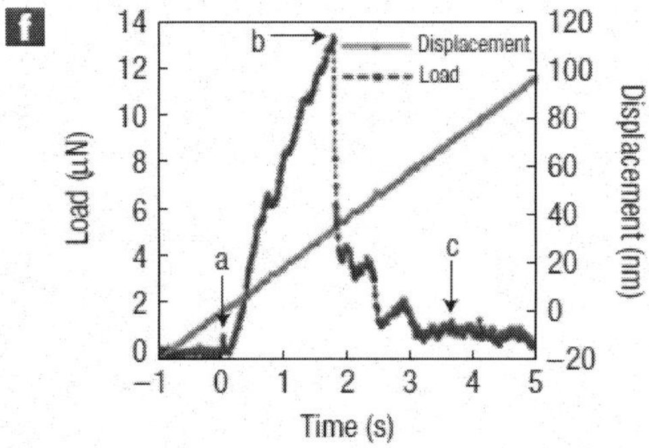

Figure 7.6 (a) through (e) TEM images of fracture of CdS nano-particle under compression. (f) Load-time and displacement-time relationships during testing. Reprinted by permission from Macmillan Publishers Ltd: Shan, Z. W., et al. (2008) "Ultrahigh stress and strain in hierarchically structured hollow nanoparticles," Nat. Mater., 7, pp. 947–952, copyright (2008).

7.3 MOLECULAR DYNAMICS (MD) SIMULATIONS

7.3.1 Motion of Atoms

In solid-state materials, atoms are closely packed at an average distance of a few 10^{-10} m (= 1 Å), while the radius of an atomic nucleus (almost 10^{-14}~10^{-15} m) is much smaller than the distance between the atoms. Since atomic nuclei are regarded as a point mass, atomic components can now successfully be modeled as a system of particles moving and interacting with other particles. Then, the motion of the gravity center of atoms can be precisely described by Newton's equation of motion based on classical dynamics.

In the framework of molecular dynamics, the motion of an atom α, follows

$$m^{\alpha}\mathbf{a}^{\alpha} = \mathbf{F}^{\alpha} \tag{7.1}$$

$$\mathbf{a}^{\alpha} = \frac{d\mathbf{v}^{\alpha}}{dt} = \frac{d^2\mathbf{r}^{\alpha}}{dt^2} \tag{7.2}$$

where t, $\mathbf{r}^{\alpha}$, $\mathbf{v}^{\alpha}$, $\mathbf{a}^{\alpha}$, and m^{α} denote the time, position, velocity, acceleration vectors, and mass of an atom α, respectively.

$\mathbf{F}^\alpha$ represents the force acting on an atom, α, which is evaluated from the interatomic potentials described below.

By numerically solving the equations of motion, Eqs. (7.1) and (7.2), we can pursue the change in the atomic arrangement. Verlet's algorithm [Verlet, 1967] is one of the simplest approaches to solving the equation of motion. Let us consider the position vector of an atom, α, at time, t, as $\mathbf{r}^\alpha(t)$. The atomic position at the time elapsing by the infinitesimal time, Δt, from the current time, t, $\mathbf{r}^\alpha(t + \Delta t)$, can be written using Taylor's expansion and neglecting any terms over the fourth order:

$$\mathbf{r}^\alpha(t+\Delta t) = 2\mathbf{r}^\alpha(t) - \mathbf{r}^\alpha(t-\Delta t) + \frac{\mathbf{F}^\alpha(t)}{m^\alpha}(\Delta t)^2 \tag{7.3}$$

Eq. (7.3) means that an unknown atomic position at a future time, $t + \Delta t$, can be evaluated from atomic positions at a past time, $t - \Delta t$, and the current time, t. Once all the atomic coordinates at the time, $t + \Delta t$, $\mathbf{r}^\alpha(t + \Delta t)$ are obtained, $\mathbf{r}^\alpha(t + 2\Delta t)$ can be calculated in the same manner. By numerically repeating the above-mentioned procedure, we can analyze the time evolution of an atomic configuration. Note that the increment of time, Δt, should be taken to be sufficiently small so that Eq. (7.3) is approved as a good approximation on the integration of Eq. (7.1). In most cases, Δt is taken to be femtosecond-order (1 fs = 1×10^{-15} s). Other algorithms for numerical integration with different accuracy and computational load have been proposed, e.g., the leapfrog finite difference algorithm, the predictor-corrector algorithm, and Gear's algorithm [Gear, 1971]. The reader should carefully choose an algorithm in the calculation that suits the simulation's purpose.

Researchers may be interested in a material's properties or atomic behavior at a specific temperature and/or under pressure. For this purpose, various methods controlling temperature and pressure have been proposed. The simplest algorithms for temperature control are found in the velocity scaling method, where the temperature in a system is adjusted to the targeted magnitude by coercively accelerating or decelerating the velocity of atoms. On the other hand, Nose's algorithm [Nose, 1984] keeps the temperature of an atomic system constant in terms of the statistical thermodynamics by contacting a super particle representing the heat bath with the system. Pressure of a system can be controlled by changing the volume of the unit cell. In Anderson's algorithm [Andersen, 1980], the simulation cell automatically expands or shrinks to keep pressure constant in terms of the statistical thermodynamics by assuming an imaginary piston on the cell. The Parinello-Rahman algorithm [Parrinello and Rahman, 1981] is an extension of Anderson's algorithm whereby increasing degrees of freedom control the stress. Simulations under constant temperature and/or constant pressure can be realized by combining these algorithms.

7.3.2 Description of Interatomic Potentials

Table 7.1 Interatomic potentials

Classification	Representatives	Precision	Cost
Empirical potential	Morse, Lennard-Jones (L-J), Born-Mayer	Low	Light
Semi-empirical potential	Metallic materials: Embedded-atom method (EAM) [Daw and Baskes, 1984], modified EAM (MEAM) [Baskes, 1992], Finnis-Sinclair (FS) [Finnis and Sinclair, 1984] Covalent materials: Stillinger-Weber (SW) [Stillinger and Weber, 1985], Tersoff [Tersoff, 1988a; Tersoff, 1988b], Brenner [Brenner, 1990], tight-binding [Wang et al., 1989] Ionic materials: Rigid-ion model, shell model [Dick and Overhauser, 1958]	Medium	Medium
Nonempirical approach	*Ab initio* (first-principles) calculations [Hohenberg and Kohn, 1964; Kohn and Sham, 1965]	High	Heavy

In a molecular dynamics simulation, the interaction of atoms is represented by the force, $\mathbf{F}^{\alpha}$, in Eq. (7.3). In other words, the reliability of simulations mostly depends on the accuracy of the interatomic interaction. In general, the interatomic potential that is a function of the atomic position describes the interaction by considering the bonding characteristics (e.g., metallic, covalent, or ionic).

Interatomic potentials are usually categorized into three types by their approximation: (1) empirical potential, (2) semi-empirical potential, and (3) non-empirical approach. The representative potentials in each group are summarized in Table 7.1.

7.3.2.1 Empirical potentials

Interatomic interaction is represented by a simple differentiable function of the distance between atoms, and the parameters in the function are fully determined empirically to reproduce the experimental results of interest, e.g., lattice constants, elastic constants, or phonon dispersion relations.

The simplest interatomic interaction is a two-body potential, where the interaction of atoms is described as a function of distance between a pair of atoms. Here, the potential energy of an atomic system, E_{tot}, can be given as

$$E_{tot} = \frac{1}{2}\sum_{\alpha}\sum_{\beta(\neq\alpha)} \phi(r^{\alpha\beta}) \tag{7.4}$$

where $r^{\alpha\beta}$ denotes the distance between atoms α and β. The Morse potential, Lennard-Jones potential, and Born-Mayer potential are well known representatives of the two-body empirical potentials. The advantage of two-body empirical potentials is that they are computationally efficient for large-scale molecular dynamics simulations because their description of potential energy is simple; the disadvantage is their low accuracy. Hence, two-body potentials are suitable for evaluating the properties of a large-scale atomic system. The two-body assumption fails to describe the anisotropy of elastic constants in a cubic crystal (Cauchy relation: C_{12}=C_{44}) and cannot distinguish binding energy from vacancy formation energy. To overcome these limitations, semi-empirical potentials have been proposed for many-body interactions.

7.3.2.2 Semi-empirical potentials

Since semi-empirical potential is more accurate than empirical potential and has low or medium computational cost, it is more likely to be used for a molecular dynamics study. Because the framework of semi-empirical potential is partially based on quantum mechanics, the functional form of semi-empirical potential includes many-body interaction, while some of parameters are determined empirically. Various models have been proposed to describe the bonding types, e.g., metallic, covalent, and ionic systems.

The embedded-atom method (EAM) proposed by Daw and Baskes, [Daw and Baskes, 1984] can successfully describe the interaction in a metallic system. In the framework of EAM, a metal atom is considered as a positively charged ion embedded in an electron cloud. The potential energy, E_{tot}, is given as

$$E_{\text{tot}} = \sum_{\alpha} F(\rho^{\alpha}) + \frac{1}{2}\sum_{\alpha}\sum_{\beta(\neq\alpha)} V(r^{\alpha\beta}) \tag{7.5}$$

where $F(\rho^{\alpha})$ denotes the embedded energy of the atom, α, embedded into an electron density of ρ^{α}. ρ^{α} is calculated by the simple summation of the charge density of atom β neighboring atom α:

$$\rho^{\alpha} = \sum_{\beta(\neq\alpha)} \rho\,(r^{\alpha\beta}) \tag{7.6}$$

One remarkable feature of EAM is its excellent transferability taking into account the environmental electronic effect in a metallic system, even though the computational load is comparable to that of a two-body empirical potential.

Finnis and Sinclair proposed another many-body potential called the FS potential [Finnis and Sinclair, 1984]. Surprisingly, the FS potential, which was derived from the tight-binding theory [Wang et al., 1989], has almost the same functional form as the EAM potential. In recent years, some extensions of the EAM potential have been proposed, e.g., the modified EAM (MEAM) [Baskes, 1992], which adds the directional term of bonds to the conventional EAM, and generalized EAM (GEAM) [Zhou et al., 2004] which can simultaneously deal with various kinds of metallic atoms in a potential function.

Materials that have covalent bonding, such as silicon and carbon, have a strong anisotropy in bonding direction due to the formation of sp3 and sp2 bonds. For this reason, the potential function for covalent materials must include the term related to the bonding angle (three-body term). The Stillinger-Weber (SW) potential [Stillinger and Weber, 1985] and the Tersoff potential [Tersoff, 1988a; Tersoff, 1988b] are two well known ones. The SW potential can successfully reproduce the bonding angle in a Si diamond structure at the equilibrium. However, the SW potential fails to reproduce the complex phase transformation in silicon due to pressure and/or temperature because it poorly describes change in the bond length and angle. On the other hand, the Tersoff potential, which is one of the bond-order potentials (BOP), is well constructed to adapt to environmental conditions (e.g., the coordination number, the bond length, and the bond angle). Because of its construction, the Tersoff potential successfully yields both the potential energy and the atomic structure in the Si diamond structure as well as graphite, bcc, and fcc structures. As an extension of the Tersoff potential, Brenner proposed a potential that can correctly reproduce a complex carbon-hydrogen interaction [Brenner, 1990].

For ionic crystals where a delicate balance between the long-range Coulomb potential and short-range repulsive potential predominate interatomic interaction, the potential energy can be represented as a linear combination of two potentials. The simplest one is the rigid-ion model, where an atom is regarded as a charged ion with a valence number. This model works well for typical ionic crystals, e.g., NaCl and MgCl; however, it poorly describes ionic polarizability under a finite external electric field. To overcome this limitation, the shell model potential [Dick and Overhauser, 1958] was proposed; in the shell model framework, an ion consists of two imaginary charged particles, a core and a shell, which mimic the atomic nuclei and electron shell, respectively. Thanks to increased degrees of freedom, the shell model potential can represent electronic polarizability by the displacement of a core and the corresponding shell. The shell model potential is often used to evaluate spontaneous polarization and complex atomic structure in the ferroelectric perovskite oxides [Shimada et al., 2008].

7.3.2.3 Non-empirical approach (first-principles calculations)

The non-empirical approach does not need to observe physical properties experimentally, but is fully based on quantum mechanics. First-principles (*ab initio*) calculations based on the density functional theory (DFT) [Hohenberg and Kohn, 1964; Kohn and Sham, 1965] is the most famous method, where the ground state total energy of a many-electron system under external potential from atomic nuclei, E_{tot}, is given by the functional of electron density, $\rho(\mathbf{r})$:

$$E_{\text{tot}} = \int v(\mathbf{r})\rho(\mathbf{r})d\mathbf{r} + T[\rho(\mathbf{r})] + \frac{1}{2}\iint \frac{\rho(\mathbf{r})\rho(\mathbf{r}')}{|\mathbf{r}-\mathbf{r}'|}d\mathbf{r}'d\mathbf{r} + E_{\text{xc}}[\rho(\mathbf{r})] \tag{7.7}$$

The first term represents the external potential energy from atomic nuclei, the second term denotes the kinetic energy of electrons, the third term is the Coulomb potential energy of electron-electron interaction, and the fourth term is the exchange-correlation energy of many-electron interaction. Since the electron density at the ground state is obtained by minimizing the total energy, E_{tot}, it is evaluated by self-consistently solving the Schrödinger equation within the one-electron approximation (the Kohn-Sham equation):

$$\left[-\frac{1}{2}\nabla^2 + v_{\text{eff}}(\mathbf{r})\right]\psi_i(\mathbf{r}) = \varepsilon_i\psi_i(\mathbf{r}) \tag{7.8}$$

where Ψ_i denotes the wave function of a state *i*, ε_i is the energy level of a state, *i*, and $v_{eff}(\mathbf{r})$ is the effective one-electron potential. In recent years, some projects have developed the first-principles code (free or charge), such as VASP, ABINIT, PWscf, and WIEN2k.

Not only the strength of materials but also various kinds of electronic properties (band structures, bandgaps, electronic polarization, magnetic moment, and conductivity) can be analyzed using first-principles calculations, which is a great advantage compared with empirical and semi-empirical potentials. Moreover, first-principles calculations, in principle, can deal with all the atoms (or their compounds) in the periodic table. The disadvantage is the heavy computational load that limits the number of atoms one can deal with—up to one hundred atoms now.

Of course, some specific atoms are quite difficult to analyze appropriately even by using first-principles calculations due to the difficulty of including relativistic effects, e.g., atoms with large atomic numbers and strongly correlated electron systems.

7.4 DEFINITIONS OF STRESS AND STRAIN IN THE ATOMIC SCALE

Stress and strain are the fundamental parameters indicating the mechanical strength of materials. However, it is hard to apply the concept of stress and strain to atomic components (discreteness of atoms) because stress and strain are originally defined in a continuum media. In this section, we review the definition of stress and strain in the atomic scale.

7.4.1 Definition of Strain in the Atomic Scale

Let us denote the coordinate of an atom α as r_i^α (i = x, y, z) and consider small deformation by applying a homogeneous external load. For the ideal perfect crystal such as the fcc and bcc, the strain, ε_i, is homogeneous anywhere in the system; this is called the homogeneous strain system (see Fig. 7.7). The coordinate of an atom α, r'^{α}_i, after a small deformation is given as

$$r_i^{\alpha\,'} = r_i^\alpha + \sum_j \varepsilon_{ij} r_j^\alpha \tag{7.9}$$

The evaluation of local strain often becomes complicated due to the discontinuity of atoms. For example, when an atomic system is inhomogeneous, e.g., it has defects or grain boundaries (called an inhomogeneous strain system), strain differs from location to location. In addition, atomic displacements may not be given by Eq.

(7.9) when the atomic structure in the unit cell is complex (e.g., Si) even though the unit cell is arranged in an orderly way. The local strain can be evaluated from the displacement of atoms β from the neighboring atom α, u_i^β, by the strain-displacement relationship:

$$\varepsilon_i^\alpha = \frac{\partial}{\partial i}\sum_\beta u_i^\beta,$$

$$\gamma_{ij}^\alpha = \frac{\partial}{\partial j}\sum_\beta u_i^\beta + \frac{\partial}{\partial i}\sum_\beta u_j^\beta \tag{7.10}$$

Even in this definition, it is ambiguous which atoms should be included in the evaluation of strain. For example, we can consider first or second neighbor atoms, or all atoms inside a wider cut-off radius.

For a more complicated case, such as metallic glasses and amorphous silicon, where the atomic arrangement is never orderly but almost random, the above-mentioned definition often gives too fine a resolution to allow us to obtain proper strain distribution. A more effective method for evaluating a highly-disordered system has been proposed using a weight factor decreasing the fluctuation in the strain [Nakatani et al., 1998].

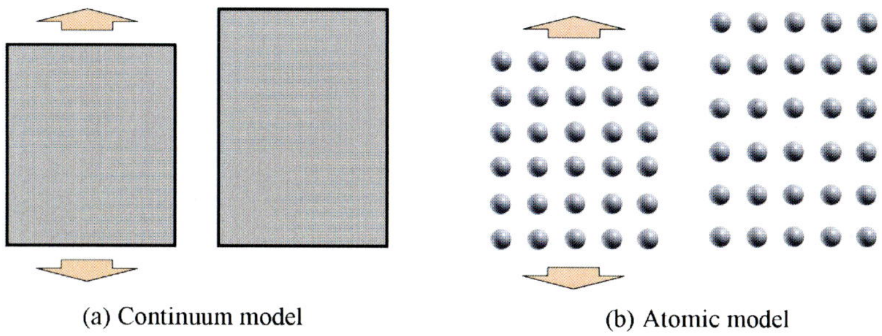

Figure 7.7 Schematic illustration explaining the concept of strain in atomic scale.

7.4.2 Definition of Stress in the Atomic Scale

7.4.2.1 Areal density of internal force

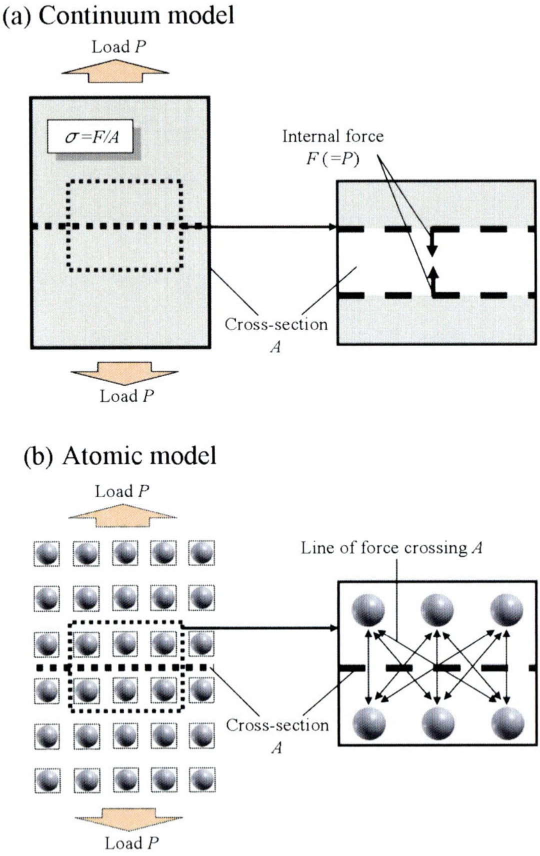

Figure 7.8 Schematic illustrations explaining the concept of stress in continuum and atomic scales.

In continuum mechanics, stress is defined as the internal force acting on a plane on an imaginary cross-section area (see Fig. 7.8a). Extending the concept to atomic systems, stress in the atomic scale is defined as the force crossing a plane within a unit area. When the interatomic interaction is described by the two-body potential, stress in the atomic scale, σ_{ij}, is given by

$$\sigma_{ij} = \frac{1}{A}\left[\sum_{S} \frac{\partial\phi(r)}{\partial r} \frac{(\hat{r}\cdot\hat{i})(\hat{r}\cdot\hat{j})}{|\hat{r}\cdot\hat{j}|}\Big|_{r=r^{\alpha\beta}}\right] \tag{7.11}$$

where the summation, Σ_S, runs over the lines of force crossing the focused plane, A (see Fig. 7.8b). We can obtain the local stress of an atom α, σ_{ij}^{α}, which is calculated from the force crossing the plane between two atoms,

$$\sigma_{ij}^{\alpha} = \frac{1}{A^{\alpha}}\left[\sum_{S'} \frac{\partial\phi(r)}{\partial r} \frac{(\hat{r}\cdot\hat{i})(\hat{r}\cdot\hat{j})}{|\hat{r}\cdot\hat{j}|}\Big|_{r=r^{\alpha\beta}}\right] \tag{7.12}$$

where the summation, $\Sigma_{S'}$, runs over all the pairs of atoms crossing the imaginary plane, A^{α}.

Since atomic stress is defined on an arbitrary plane, it is effective to evaluate the stress distribution at a grain boundary or interface. On the other hand, only the pair potential interaction is applicable.

7.4.2.2 Definition based on strain energy

In the continuum theory, stress, σ_{ij}, is defined as the first-order differential of total energy, E, in response to the strain. Since the total energy is equal to the potential energy, Φ, at zero temperature ($T = 0$ K), the stress can be given by

$$\sigma_{ij} = \frac{1}{V}\frac{\partial\Phi}{\partial\varepsilon_{ij}} \tag{7.13}$$

where V denotes the volume of the system. The advantage of the definition is that stress can always be calculated in any description of potential energy.

Although the complex many-body effect influences the potential energy of the system, it is often approximated in the simple form

$$\Phi = \sum_{\alpha}^{N} \Phi^{\alpha} \tag{7.14}$$

where Φ^{α} denotes the potential energy of an atom α. Not only the empirical two-body potentials but also the accurate many-body potentials such as the embedded atom method (EAM), the effective medium theory (EMT), and the Tersoff potentials take this form in the expression. Under the approximation, the stress in Eq. (7.13) can be written as

$$\sigma_{ij} = \frac{1}{V}\sum_{\alpha}^{N}\left(\frac{\partial \Phi^{\alpha}}{\partial \varepsilon_{ij}} + \frac{\partial \Phi^{\alpha}}{\partial \xi^{\alpha}}\frac{\partial \xi^{\alpha}}{\partial \varepsilon_{ij}}\right) = \frac{1}{V}\sum_{\alpha}^{N}\frac{\partial \Phi^{\alpha}}{\partial \varepsilon_{ij}} \tag{7.15}$$

by supposing that the local strain at the site of atom α is denoted as $\varepsilon_{ij} + \zeta^{\alpha}$, where ζ^{α} is the deviation from the averaged global strain due to the inhomogeneous deformation at the site of atom α (inhomogeneous contribution). Here, $\partial\Phi^{\alpha}/\partial\zeta^{\alpha} = 0$ because the atomic configuration is stable ($T = 0$ K). In the same manner, the local stress of an atom, α, (atomic stress), is defined as the differential of potential energy of an atom, α, in terms of the strain:

$$\sigma_{ij}^{\alpha} = \frac{N}{V}\frac{\partial \Phi^{\alpha}}{\partial \varepsilon_{ij}} \tag{7.16}$$

Substituting Eq. (7.16) into Eq. (7.15), the stress is given by the average of the atomic stress:

$$\sigma_{ij} = \frac{1}{N}\sum_{\alpha}^{N}\sigma_{ij}^{\alpha} \tag{7.17}$$

The distribution of atomic stress agrees well inside a material with a stress field obtained by the continuum theory.

This means that we can define stress in an atom-by-atom scale. Thus, the relationship between the stress concentration near a crack tip or interface edge, which was discussed in earlier chapters based on the continuum (fracture mechanics) theory, can be also investigated at the atomistic level by using atomic stress, σ_{ij}^{α}. On the other hand, there is some contradiction of the continuum theory, e.g., nonzero normal stress at a free surface and difficulty in the definition of volume at surface.

7.4.2.3 Definition of stress including the effect of temperature (Virial theorem)

According to the Virial theorem, stress under finite temperature can be derived from the balance between forces acting on the wall of a cell and external pressure. The Virial stress is given as

$$\sigma_{ij} = \frac{1}{V}\left(-\sum_{\alpha} \frac{p_i^\alpha p_j^\alpha}{m^\alpha} - \sum_{\alpha} F_i^\alpha r_j^\alpha \right) \tag{7.18}$$

where the first and second terms of the right-hand member denote the contribution of kinetic energy and that of potential energy, respectively. Here, F_i^α and p_i^α represent the force and the kinetic momentum of an atom α, in the i direction. Since each term can be calculated from the kinetic momentum and atomic position at the time, the stress is independent of simulation conditions such as constant energy, constant temperature, and so on. Note that another definition based on the thermodynamics theory gives the same expression of stress [Wallace, 1972].

Virial atomic stress can also be defined as

$$\sigma_{ij}^\alpha = \frac{N}{V}\left(-\frac{p_i^\alpha p_j^\alpha}{m^\alpha} - F_i^\alpha r_j^\alpha \right) \tag{7.19}$$

and Virial stress can be expressed as the average of the atomic stress:

$$\sigma_{ij}^\alpha = \frac{1}{N} \sum_{\alpha}^{N} \sigma_{ij}^\alpha \tag{7.20}$$

Since Virial (atomic) stress fluctuates over time under a finite temperature, the stress distribution should be normally evaluated by averaging the stress for a long period after the system reaches the thermodynamic equilibrium.

7.5 DEFINITION OF ELASTIC COEFFICIENT IN THE ATOMIC SCALE

According to the linear-elastic theory, the relationship between stress and strain is assumed to be linear,

$$\sigma_{ij} = C_{ijkl}\varepsilon_{kl} \tag{7.21}$$

where C_{ijkl} denotes the elastic coefficient. The elastic coefficient in the atomic component, C_{ijkl}, is given by the differential of stress in response to the strain:

$$C_{ijkl} = \frac{\partial \sigma_{ij}}{\partial \varepsilon_{kl}} = \frac{1}{V}\frac{\partial^2 E}{\partial \varepsilon_{ij}\partial \varepsilon_{kl}} \tag{7.22}$$

Under $T = 0$ K, the elastic constant is written using Eq. (7.14) as

$$C_{ijkl} = C^0_{ijkl} + C^{'}_{ijkl}, \tag{7.23}$$

$$C^0_{ijkl} = \frac{1}{V}\sum_{\alpha}^{N}\left(\frac{\partial^2 \Phi^\alpha}{\partial \varepsilon_{ij}\varepsilon_{kl}}\right) \tag{7.24}$$

$$C^{'}_{ijkl} = \frac{1}{V}\sum_{\alpha}^{N}\left(\frac{\partial^2 \Phi^\alpha}{\partial \xi^\alpha \varepsilon_{ij}}\frac{\partial \xi^\alpha}{\partial \varepsilon_{kl}} + \frac{\partial^2 \Phi^\alpha}{\partial \xi^\alpha \varepsilon_{kl}}\frac{\partial \xi^\alpha}{\partial \varepsilon_{ij}}\right) \tag{7.25}$$

It should be noted that the latter term for inhomogeneous strain contribution, C'_{ijkl}, is not zero. As in the case of stress, the elastic constant of atom α (the atomic elastic constant) can be given by

$$C^\alpha_{ijkl} = C^{0\alpha}_{ijkl} + C^{'\alpha}_{ijkl}$$
$$= \frac{N}{V}\left(\frac{\partial^2 \Phi^\alpha}{\partial \varepsilon_{ij}\varepsilon_{kl}}\right) + \frac{N}{V}\left(\frac{\partial^2 \Phi^\alpha}{\partial \xi^\alpha \varepsilon_{ij}}\frac{\partial \xi^\alpha}{\partial \varepsilon_{kl}} + \frac{\partial^2 \Phi^\alpha}{\partial \xi^\alpha \varepsilon_{kl}}\frac{\partial \xi^\alpha}{\partial \varepsilon_{ij}}\right) \tag{7.26}$$

Since it is, in general, quite difficult to derive the analytical formula of inhomogeneous contribution C'_{ijkl} or C'^{α}_{ijkl} for many-body potentials such as the EAM, only the homogeneous contribution, C^0_{ijkl} or $C^{0\alpha}_{ijkl}$ is usually discussed. Such treatment is not exact but provides a qualitative insight into local stiffness in an inhomogeneous structure, e.g., grain boundary [Kitamura et al., 2003].

Ray and Rahman [1984] derived the analytical formula of an elastic coefficient under a finite temperature, which is called the fluctuation formula. This approach evaluates the elastic coefficient by time-averaging the effect of thermal fluctuation. According to the fluctuation formula, the elastic coefficient under the finite temperature and the stress-free condition (HtN ensemble) is given by

$$\begin{aligned} C_{ijkl} &= -\frac{k_B T}{V}\left(\left\langle \varepsilon_{ij}\varepsilon_{kl}\right\rangle - \left\langle \varepsilon_{ij}\right\rangle\left\langle \varepsilon_{kl}\right\rangle\right) \\ &\quad + \frac{2Nk_B T}{V}(\delta_{il}\delta_{jk} + \delta_{ik}\delta_{jl}) \\ &\quad + \frac{1}{V}\sum_{\alpha}^{N}\left(\frac{\partial^2 \Phi^\alpha}{\partial \varepsilon_{ij}\varepsilon_{kl}}\right) \end{aligned} \tag{7.27}$$

where k_B and δ_{ij} denote the Boltzman constant and Kronecker delta, respectively. The variables in the angled brackets denote the time-averaging quantity. The first of these is the fluctuation term, meaning the influence of thermal fluctuation. The second is the kinetic term, representing the contribution of the kinetic energy of the atoms. The third term, the Born term, exhibits the contribution of potential energy. Note that, in the case of low temperature, long-time simulations are necessary to correctly evaluate the fluctuation term due to slight thermal fluctuations.

7.6 STRENGTH OF ATOMIC COMPONENTS

7.6.1 Ideal Strength

Ideal strength (theoretical strength) is defined by Milstein as the stress or strain at which a perfect crystal lattice becomes mechanically unstable under arbitrary homogeneous deformation [Milstein, 1973]. Ideal strength has been intensely investigated because the ultimate strength without defects is a fundamental mechanical characteristic of materials. In other words, ideal strength physically means the maximum resistance against external stress (or strain) that a material can possess. However, it is well known that there is an important difference between ideal strength and actual strength observed experimentally, and this fact led researchers to discover dislocation, which plays an important role in crystal plasticity. On the other hand, such high strength is experimentally observed in whiskers that include few defects. Recently, the strength obtained by nano-indentation is attracting researchers' attention because extremely localized evaluation reveals properties in a defect-free region [Bahr et al., 1998; Woodcock and Bahr, 2000; Gouldstone et al., 2000]. As we will mention again later, the fact that few defects can stably exist in a nanoscale material suggests that ideal strength can become actual strength in such nanoscale materials.

To investigate the mechanical behavior of nano-components, it is essential to analyze their ideal strength from the atomistic (or more precisely, electronic) level. The stress and strain curve of a perfect crystal is obtained by applying periodic boundary conditions to the unit cell (realizing a perfect crystal) under homogeneous strain. The first-principles calculation with the plane-wave basis set is suitable for such a simulation. Since the first investigation of the tensile ideal strength of Cu by Esposito et al. [1980], a number of studies on ideal tensile/shear strengths of various kinds of materials have been done, which are listed in Table 7.2. For example, the total energy and stress of iron as a function of strain obtained from first-principles calculations based on the accurate FLAPW method by Friák et al. [2003] are shown in Fig. 7.9. The stress increases with the strain, and it drops dramatically after the stress reaches its maximum. The peak of stress denotes the ultimate strength of iron, namely, ideal strength.

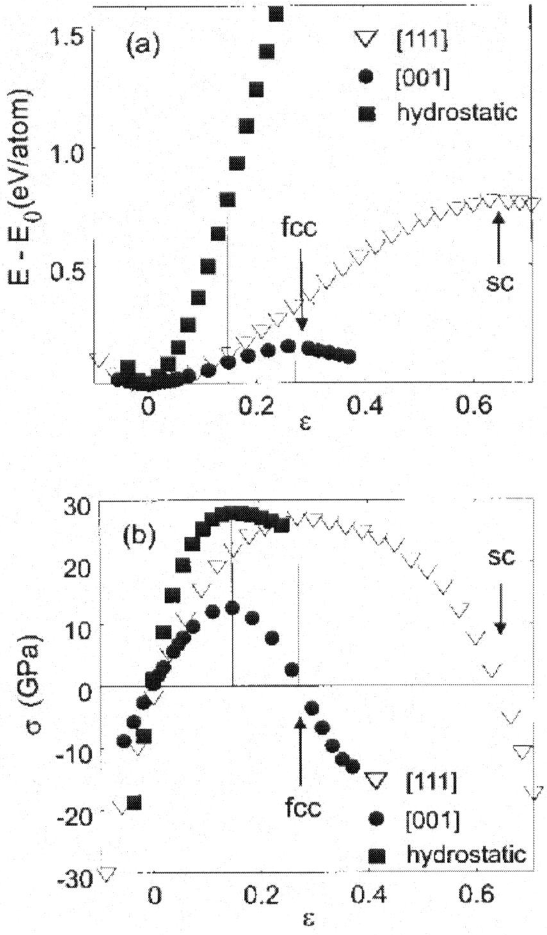

Figure 7.9 Energy and stress as a function of strain of Fe.Reprinted Šob, M., et al. (2004) "The role of ab initio electronic structure calculations in studies of the strength of materials," *Mater. Sci. Eng. A*, 387–389, pp. 148–157, Copyright (2004), with permission from Elsevier.

Table 7.2 Ideal theoretical shear and tensile strength of various kinds of materials

Crystal	Structure	Shear direction/ plane	Theoretical strength
Li	bcc	111/{110}	0.319 [Sandera and Pokluda, 1994]
Na	bcc	111/{110}	0.209 [Sandera and Pokluda, 1993]
		111/{112}	0.20 [Morris et al., 2000]
K	bcc	111/{110}	0.112 [Sandera and Pokluda, 1993]
Ge	dia	110/{111}	12.0 [Sandera and Pokluda, 1993]
Si	dia	110/{111}	14.7 [Sandera and Pokluda, 1993]
C	dia	110/{111}	131 [Sandera and Pokluda, 1993]
Al	fcc	112/{111}	0.75 [Macmillan, 1983]
		112/{111}	0.9 [Kelly and Macmillan, 1986]
		110/{111}	2.62 [Macmillan, 1983]
		112/{111}	2.56 [Sandera and Pokluda 1993]
		112/{111}	3.50 [Morris et al., 2000]
		112/{111}	3.40 [Morris et al., 2000]
		112/{111}	1.85 [Morris et al., 2000]
Cu	fcc	112/{111}	1.1 [Macmillan, 1983]
		112/{111}	1.29 [Kelly and Macmillan, 1986]
		112/{111}	1.20 [Kelly and Macmillan, 1986]
		112/{111}	2.93 [Sandera and Pokluda, 1993]
		112/{111}	5.30 [Morris et al., 2000]
		112/{111}	4.00 [Morris et al., 2000]
		112/{111}	2.65 [Morris et al., 2000]
		011/{011}	0.80 [Kelly and Macmillan, 1986]
		110/{111}	0.61 [Kelly and Macmillan, 1986]
		110/{111}	0.46 [Kelly and Macmillan, 1986]
Ag	fcc	112/{111}	0.66 [Kelly and Macmillan, 1986]
		110/{111}	0.77 [Kelly and Macmillan, 1986]
		112/{111}	1.90 [Sandera and Pokluda,1993]
		011/{111}	0.71 [Kelly and Macmillan, 1986]
Fe	bcc	111/{110}	7.3 [Macmillan, 1983]
		111/{110}	6.6 [Kelly and Macmillan, 1986]
		111/{110}	11.5 [Sandera and Pokluda, 1993]
		111/{011}	3.56 [Kelly and Macmillan, 1986]
V	bcc	111/{112}	7.3 [Morris et al., 2000]
		111/{110}	9.72 [Sandera and Pokluda, 1994]
Cr	bcc	111/{112}	20.5 [Morris et al., 2000]

Crystal	**Structure**	**Shear direction/ plane**	**Theoretical strength**
Ni	fcc	112/{111}	5.79 [Sandera and Pokluda, 1994]
		010/{100}	9.1 [Macmillan, 1983]
		112/{111}	2.4 [Kelly and Macmillan, 1986]
		110/{111}	7.1 [Kelly and Macmillan, 1986]
It	fcc	110/{111}	18.6 [Sandera and Pokluda, 1994]
		112/{111}	24.1 [Morris et al., 2000]
Pb	fcc	112/{111}	0.472 [Sandera and Pokluda, 1994]
Mo	bcc	111/{112}	17.8 [Morris et al., 2000]
		111/{112}	20.9 [Morris et al., 2000]
Ar	fcc	110/{110}	0.14 [Sandera and Pokluda, 1993]
Crystal	**Structure**	**Tensile direction**	**Theoretical strength**
Na	bcc	100	0.036 [Macmillan, 1983]
		110	0.191 [Macmillan, 1983]
		111	0.848 [Macmillan, 1983]
Ge	dia	100	15.2 [Macmillan, 1983]
		111	37.1 [Macmillan, 1983]
		100	15.2 [Sandera and Pokluda, 1994]
Si	dia	100	22.4 [Macmillan, 1983]
		111	47.3 [Macmillan, 1983]
		111	32 [Macmillan, 1983]
		111	24.8 [Huang et al., 1994]
		111	32.0 [Sandera and Pokluda, 1994]
		100	22.4 [Sandera and Pokluda, 1994]
		111	22.0 [Roundy and Cohen, 2001]
C	dia	100	168 [Macmillan, 1983]
		111	205 [Macmillan, 1983]
		111	205 [Sandera and Pokluda, 1994]
		110	168 [Sandera and Pokluda, 1994]
		111	90 [Telling et al., 2000]
		111	95 [Roundy and Cohen, 2001]
		100	225 [Telling et al., 2000]]
		100	130 [Telling et al., 2000]
Al	fcc	100	12.1 [Roundy et al., 2001]
		111	11.1 [Li and Wang, 1998]
		100	3.49 [Sandera and Pokluda, 1994]
		111	11.0 [Ogata and Kitagawa, 1999]
		100	13.1 [Kitamura and Umeno, 2003]
		100	9.2 [Clatterbuck et al., 2003]
		111	8.95 [Clatterbuck et al., 2003]
		110	4.89 [Clatterbuck et al., 2003]

Crystal	Structure	Tensile direction	Theoretical strength
Cu	fcc	100	32.0 [Esposito et al., 1980]
		100	41.0 [Esposito et al., 1980]
		111	39.0 [Macmillan, 1983]
		001	33.0 [Sob et al., 1998]
		110	31.0 [Sob et al., 1998]
		001	9.3 [Cerny et al., 2004]
Ag	fcc	111	24 [Macmillan, 1983]
Au	fcc	110	27.0 [Macmillan, 1983]
		111	27.0 [Sandera and Pokluda, 1994]
Fe	bcc	111	47.9 [Sandera and Pokluda, 1994]
		100	30.0 [Sandera and Pokluda, 1994]
		001	12.7 [Friák et al., 2003]
		001	14.2 [Clatterbuck et al., 2001]
Mo	bcc	001	28.8 [Luo et al., 2002]
W	bcc	001	28.9 [Sob et al., 1997]
		111	40.1 [Sob et al., 1997]
		100	90.8 [Macmillan, 1983]
Ar	fcc	100	0.32 [Sandera and Pokluda, 1994]

7.6.2 Ideal Strength of Low-Dimensional Nano-materials

In recent years, materials on a nanometer scale, e.g., ultrathin films, nanowires, and atomic chains, have been manufactured not only out of scientific interest but also in anticipation of future practical applications in electronic devices. Such small materials exhibit different characteristic strength than their macroscale counterparts. Figure 7.10 shows the yield strength of Au with the size from submicron to atomic scale obtained both experimentally and theoretically [Greer, 2006]. As the size of the Au specimen decreases, the yield strength increases rapidly and it approaches ideal strength. Because the nanoscale materials contain few defects such as dislocation, suggesting an almost perfect crystal, the yield strength can become close to the ideal strength.

It is well known that the strength of materials greatly depends on the structure. Picture, for example, component shapes: films, wires, tubes, and dots. When we look into the component, we find an orderly arrangement of atoms, e.g., a face-centered cubic (fcc) or body-centered cubic (bcc) crystallographic structure, and the local disorder of atoms such as a grain boundary and dislocation. Such an atomic arrangement can be categorized as an understructure, while the component shape is regarded as an external structure (global geometry).

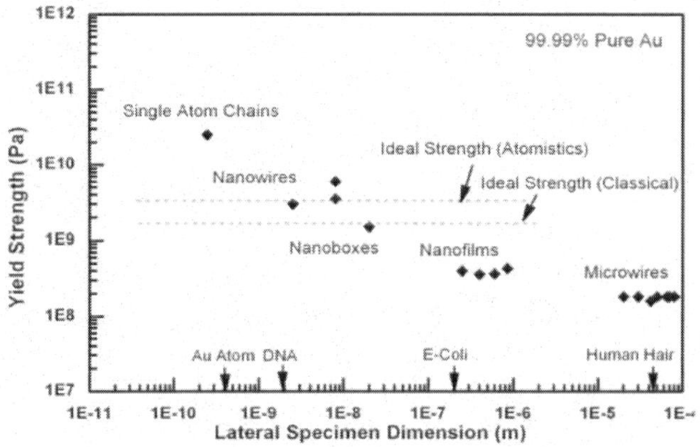

Figure 7.10 Experimental measurement and theoretical investigation of the yield strength of Au nano-structures as a function of lateral specimen dimension. Reprinted from Greer, J. R. (2006) "Bridging the gap between computational and experimental length scales: a review on nanoscale plasticity," *Rev. Adv. Mater. Sci.*, 13, pp. 59–70, with permission from the Russian Academy of Sciences.

Let us consider the effect of the external and internal structure on strength using a simple example of macro- and nano-scale materials. Figure 7.11a shows a macroscale tube of steel in a power plant with a diameter and length of a centimeter and meter order, respectively. The tube consists of a number of grains, and the grain size is normally in the order of tens of micrometers. Of course, in this macroscale case, the tube corresponds to the external structure, and the smaller grains are the understructure. Now consider carbon nanotubes (CNTs), which are made by rolling up a graphene sheet in a specific direction, and have a similar tubal shape (see Fig. 7.11b). The same external structure (shape) as the macroscale tube can be realized even in the nanoscale. Like the macroscale tube, the CNT has an understructure, namely a hexagonal honeycomb lattice (the size of several angstroms). The important feature is that the size is extremely close to that of the external structure (the size of several nanometers), as shown in Fig. 7.11b, while the scale differs largely (10^{-6}~10^{-3} order) in the macroscale material (Fig. 7.11a). This signifies that it is essential to investigate simultaneously the influence of both the external and internal structures when we study the strength of nanoscale components.

The nano-components characteristically have a low-dimensional periodic structure, such as nano-films (two-dimensional; 2D), nanowires and nanotubes (one-dimensional; 1D), and nano-dots (zero-dimensional; 0D). Considering that nano-components have few defects, as mentioned above, they have a clean and defect-free structure, in other words, high periodicity. We call a low-dimensional and highly periodic structure an ideal structure. To understand the strength of nano-components, it is essential we pick a typical ideal structure and investigate its strength from external and internal structural points of view.

In the following sections, we review some of the research on ideal strength that the typical ideal structure intrinsically possesses in terms of both its external and internal structures.

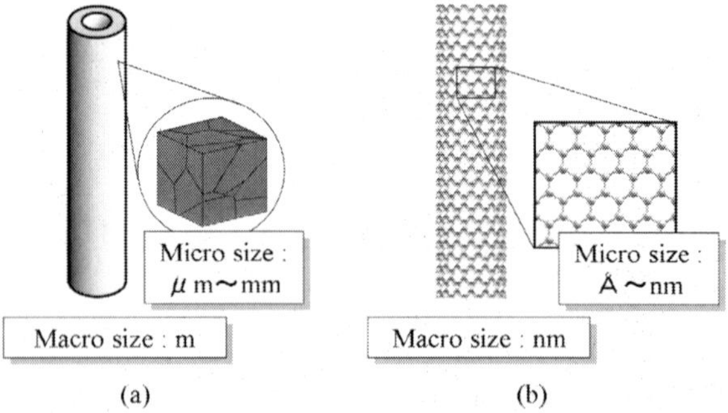

Figure 7.11 Comparison of macroscopic and microscopic structures in the (a) macroscale iron tube and its grains, and (b) the nanoscale carbon nanotube (CNT) and its atomic configuration.

7.6.2.1 Ideal strength of Si nano-films (2D)

On a Si (001) surface, there exists a drastic surface reconstruction by a pair of dangling bonds (a dimer). The strength of Si nano-film must be affected not only by surfaces as an external structure but also by local change in the atomic arrangement (surface reconstruction) as understructure.

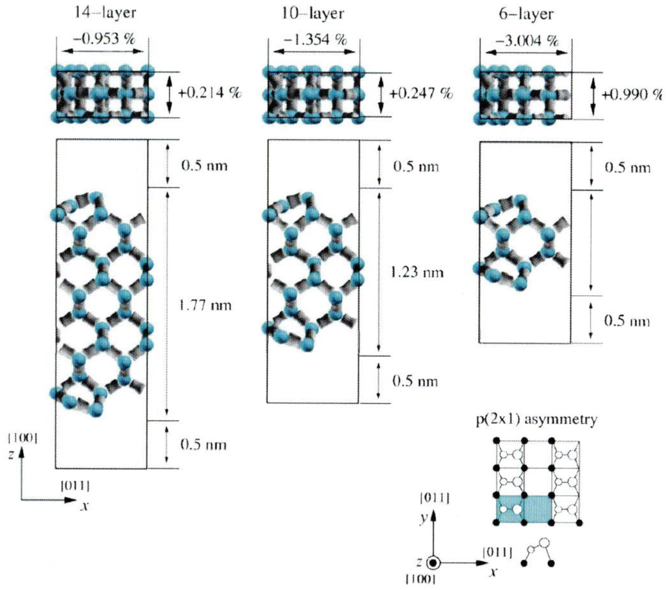

Figure 7.12 Si nano-film of thickness of 14, 10, and 6 atomic layers with the p(2×1) reconstructed surfaces. The tensile axis lies along the [011] direction. Reprinted with permission from Umeno, Y., et al. (2005) "*Ab initio* study of the surface properties and ideal strength of (100) silicon thin films," *Phys. Rev. B*, 72, article number 165431. Copyright (2005) by the American Physical Society.

Figure 7.12 displays the simulation models of Si nano-films with a (100) surface having $p(2 \times 1)$ asymmetric dimers [Umeno et al., 2005]. The Si nano-films consisting of 14, 10, and 6 atomic layers are prepared to investigate the influence of thickness. Figure 7.13 plots the stress-strain curves under [110] tension obtained by first-principles calculation with ultrasoft pseudopotentials (USPP)

[Vanderbilt, 1990] within the GGA [Perdew and Wang, 1992], comparing with the result of bulk. Under a small strain ($\varepsilon < 0.05$), the Young's modulus of nano-films is almost same as that of bulk. This is because the number of covalent bonds per a cross-section area is almost the same. Considering the fact that electrons in a covalent material are highly localized at the bonding site among atoms, surface is unlikely to affect the strength of covalent bonds inside nano-films. On the other hand, the ideal strength (the maximum stress of the stress-strain relationship) exhibits a clear difference; the ideal strengths of 10- and 6-layered Si nano-films are weaker, compared with that of bulk, while the thicker, 14-layered film possesses almost the same strength as bulk.

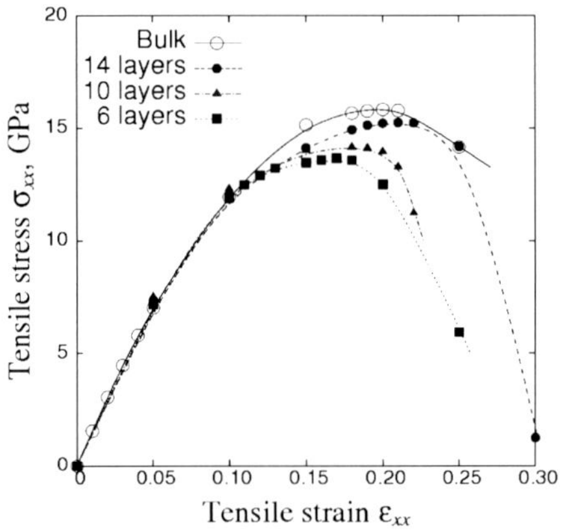

Figure 7.13 Stress and strain curves of Si nano-films and bulk under [011] tension (*ab initio* calculations with the ultrasoft pseudopotential within the GGA). Reprinted with permission from Umeno, Y., et al. (2005) "*Ab initio* study of the surface properties and ideal strength of (100) silicon thin films," *Phys. Rev. B*, 72, article number 165431. Copyright (2005) by the American Physical Society.

Figure 7.14 shows the change in atomic configuration and isosurface of high charge density (0.425×10^3 nm^{-3}) during [110] tension. As the tensile strain is applied, the covalent bond shown as the isosurface shrinks due to deceasing charge density, and finally the bond partially breaks. It should be noted that we can reasonably regard this as a break of bond when we consider the drastic drop in stress, even though actually there exists a slight charge density at the broken site. For the 14-layered nano-film, the covalent bonds inside the film support the entire structure after the bonds α and β at the surface break. In addition, the atomic structure inside the film deforms as homogeneously as bulk. For the 10- and 6-layered nano-films, the breakdown of the α and β surface bonds develops inside along the dashed line, which causes inhomogeneous fracture. Thus, such change in fracture process from the homogeneous and entire fracture (bulk and 14-layered film) to the inhomogeneous and localized fracture (10- and 6-layered films) due to the reconstructed surface structure gives rise to the weakening in ideal strength.

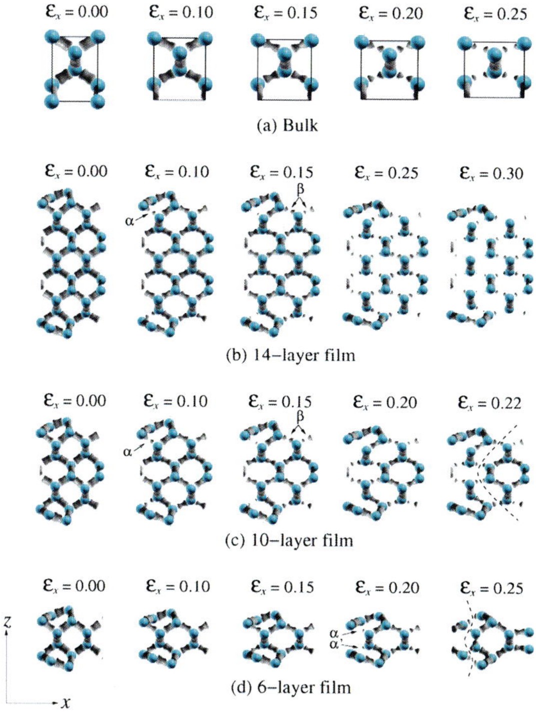

Figure 7.14 Change in the atomic configuration and covalent bond of Si nano-films and bulk under [011] tension (the same calculation as Fig. 7.10). Blue spheres are Si atoms and gray regions indicate iso-surface of charge density of 0.425×10^3 nm^{-3}. Reprinted with permission from Umeno, Y., et al. (2005) "*Ab initio* study of the surface properties and ideal strength of (100) silicon thin films," *Phys. Rev. B*, 72, article number 165431. Copyright (2005) by the American Physical Society.

7.6.2.2 Ideal strength of Cu nano-films (2D)

Nano-film is one of the types of two-dimensional nano-components. Surface structure is expected to significantly affect the strength of nano-films because the ratio of surface with respect to entire volume is quite high. On the defect-free clean Cu (111) surface, there is no surface reconstruction as well as little relaxation of atomic position. Hence, the Cu nano-film is a typical example to extract the fundamental influence of surfaces as a purely external structure on strength because there is little disorder of atomic arrangement at the surface (the absence of change in the understructure).

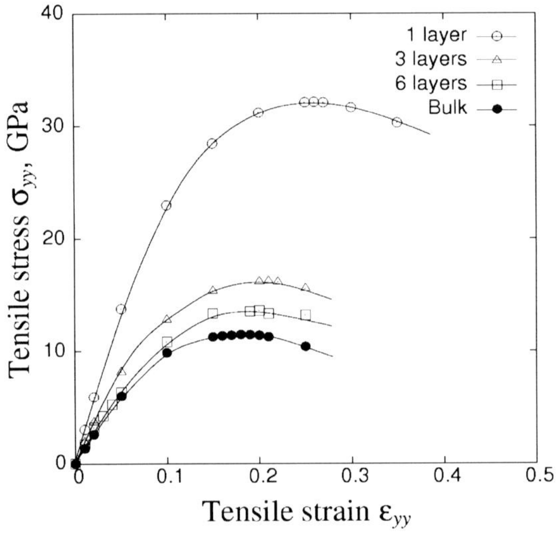

Figure 7.15 Stress-strain curves of 1, 3, 6 atomic layers of Cu slabs with (111) surfaces and Cu bulk under [112] tension (*ab initio* calculations with the ultrasoft pseudopotential within the GGA). Reprinted with permission from Kushima, A. (2007) "*Ab initio* studies on ideal strength of low-dimensional structures," PhD Thesis, Kyoto University, pp. 39–46.

Figure 7.15 shows the stress-strain curves of Cu nano-films with the (111) surface under [112] tension [Kushima, 2007]. To investigate the effect of thickness, 1-, 3-, and 6-layered Cu nano-films and Cu bulk were analyzed as well. Here, the ultrasoft pseudopotentials (USPP) within the generalized gradient approximation (GGA) were employed. At the small strain region ($\varepsilon < 0.05$), nano-films possess larger Young's modulus as the thickness of the film becomes smaller. As the strain is applied, the inclination of stress-strain curves decreases. The maximum peak stress (the ideal strength of Cu nano-films) is considerably higher than that of bulk. As the thickness of the film becomes smaller, nano-films exhibit higher ideal strength.

Figure 7.16 shows the difference of charge density distribution between the 3-layered nano-film and the linear combination of monoatomic-layered nano-film arranged in each atomic layer of 3-layered nano-film. The difference in valence-charge distribution reveals that electrons transfer from the in-plane site to the interlayer site to connect the layers. This implies that the tensile strength along the in-plane direction is weakened because the electrons contributing to the in-plane bonding decreases due to the charge transfer. Such an electron rearrangement can be regarded as the effect of understructure as the definition of understructure further scales down from the atomic level to the electron level.

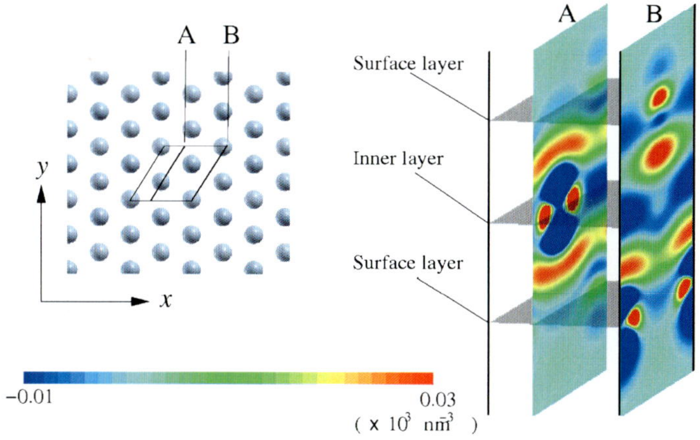

Figure 7.16 Difference in charge density distribution between the combined three Cu sheets and 3-layered Cu nano-film (the same calculation as Fig. 7.12). Electrons migrate from the in-plane site to the interlayer site. Reprinted with permission from Kushima, A. (2007) "*Ab initio* studies on ideal strength of low-dimensional structures," PhD Thesis, Kyoto University, pp. 39–46.

7.6.2.3 Ideal strength of Cu nanowires and atomic chain (1D)

Surface strongly affects the ideal strength of one-dimensional (1D) nano-structures such as nanowires, nanotubes, and atomic chains, which is the same feature as that of 2D nano-structures. In addition, nanowires often possess a characteristic atomic arrangement that totally differs from that of bulk. An atomic chain where atoms are arranged in a (straight or zigzag) line is one of the well-known examples. In addition, metallic nanowires (e.g., Au and Cu) with an extremely small diameter (down to several nanometers) has a multi-shell atomic structure [Kondo and Takayanagi, 2000], which is never the original fcc structure. Thus, the influence of the microscopic atomic arrangement is remarkably dominant in nanowires. In this section, we review theoretical studies on the mechanical strength of a Cu atomic chain and Cu multi-shell nanowires.

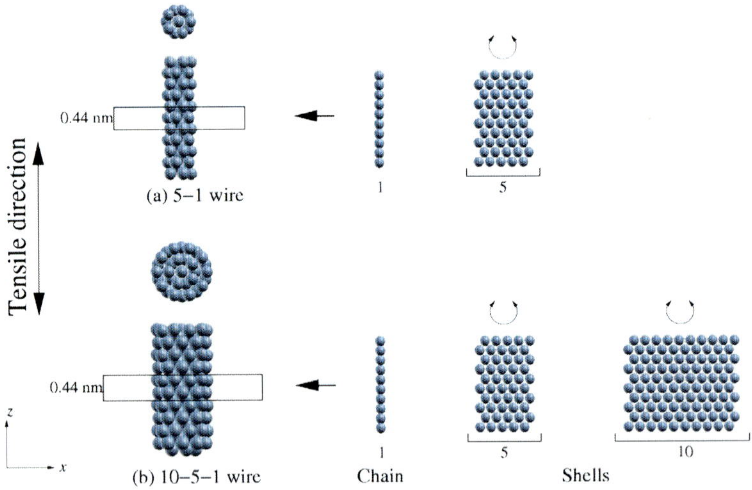

Figure 7.17 Atomic structures of multi-shell Cu nanowires of the 5-1 and 10-5-1 types. Reprinted from Kushima, A., et al. (2006) "Ideal strength of a Cu multi-shell nano-wire," *Mod. Simul. Mater. Sci. Eng.*, 14, pp. 1031–1039, with permission from IOP Publishing.

Figure 7.17 depicts the simulation models of a multi-shell Cu nanowire that consists of an atomic chain at the center (core) and a curled (111) atomic sheet (shell) [Kushima et al., 2006]. This structure, which is absolutely different from the intrinsic fcc crystal structure, was observed experimentally.

Figure 7.18 shows the stress-strain curves of a Cu atomic chain and multi-shell nanowires (an (111) atomic sheet and bulk fcc for comparison) under tension, obtained by first-principles calculations with the USPP within the GGA. The atomic chain exhibits the highest ideal strength. From the charge density distribution in Fig. 7.19, we can see that the electrons are localized among the neighboring Cu atoms and the number of electrons contributing to a bond increases because of the reduced coordination number. This charge transfer strengthens the Cu bond, which leads to a high strength.

On the other hand, the multi-shell 5-1 nanowire exhibits a lower ideal strength than that of an atomic chain and atomic sheet. Figure 7.19 shows the difference in charge density distribution between the 5-1 multi-shell nanowire and the simple combination of free-standing atomic chain and curled sheet (isolated core and shell). The electron distribution clearly changes when the atomic chain and sheet are combined. In the 5-1 multi-shell nanowire, the charge density along the atomic chain (core) site and in the curled sheet (shell) decreases, whereas that between the core and shell increases. This suggests that electrons among atoms transfer from the axial direction to circumferential direction. Thus, the depletion of axial charge density due to electron redistribution leads to the surface bond breaking under tension (see Fig. 7.21), which weakens the ideal tensile strength.

Figure 7.22 shows the change in atomic and electronic structures of the 10-5-1 multi-shell nanowire under tension. Even in the unstrained 10-5-1 nanowire, the surface bonds are not homogeneous and there exists a locally weak bond because of the misfit between the outer (10) and middle (5) shells. The weak site is preferentially extended during tensile loading, which leads to a large gap, while the surface bond simultaneously breaks in the case of the homogeneously arranged 5-1 nanowire. This inhomogeneous deformation due to the characteristic atomic arrangement results in the weakest ideal strength.

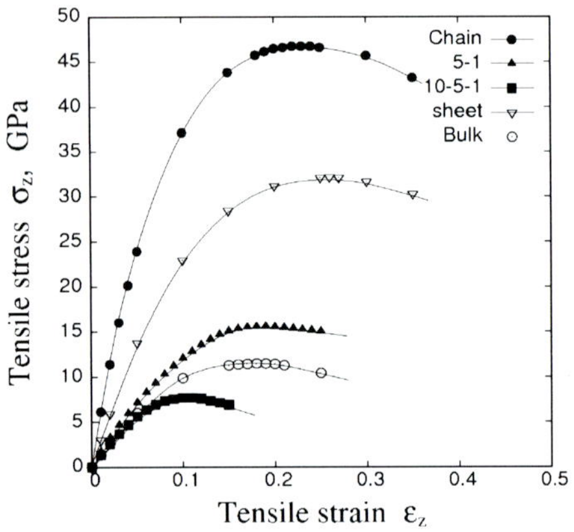

Figure 7.18 Stress and strain curves of a Cu atomic chain, atomic sheet, 5-1 and 10-5-1 multi-shell nanowires, and bulk under axial tension (*ab initio* calculations with the ultrasoft pseudopotential within the GGA). Reprinted from Kushima, A., et al. (2006) "Ideal strength of a Cu multi-shell nanowire," *Mod. Simul. Mater. Sci. Eng.*, 14, pp. 1031–1039, with permission from IOP Publishing.

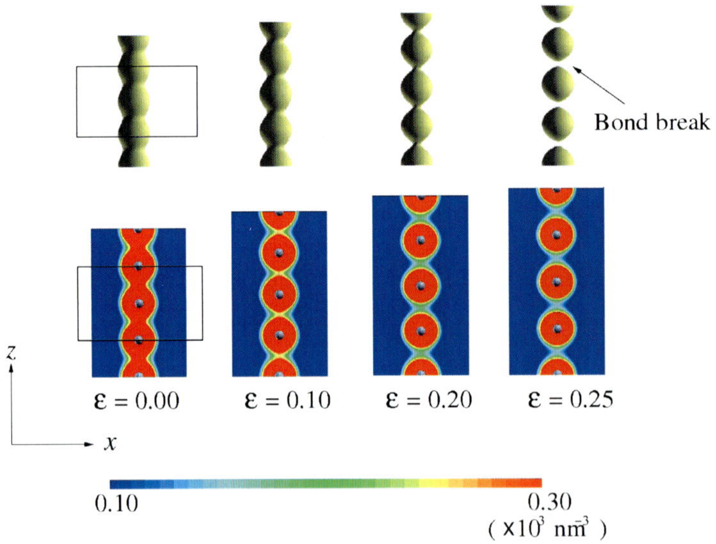

Figure 7.19 Change in atomic and electronic structures of Cu atomic chain during tensile loading (the same calculation as in Fig. 7.18).

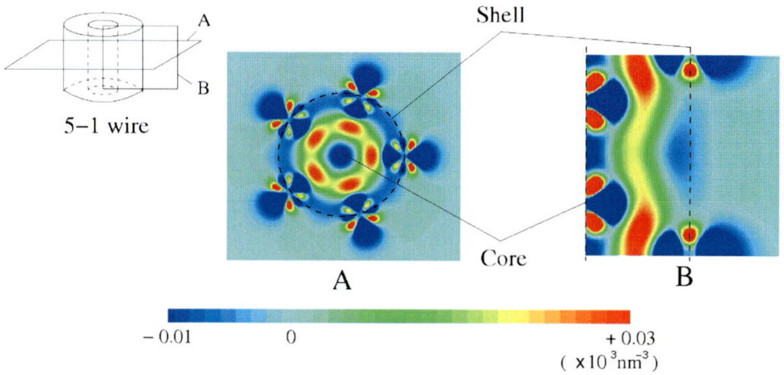

Figure 7.20 Difference in charge density distribution between a 5-1 multi-shell nanowire and combined atomic chain and curled sheet (the same calculation as in Fig. 7.18). Reprinted from Kushima, A., et al. (2006) "Ideal strength of a Cu multi-shell nano-wire," *Mod. Simul. Mater. Sci. Eng.*, 14, pp. 1031–1039, with permission from IOP Publishing.

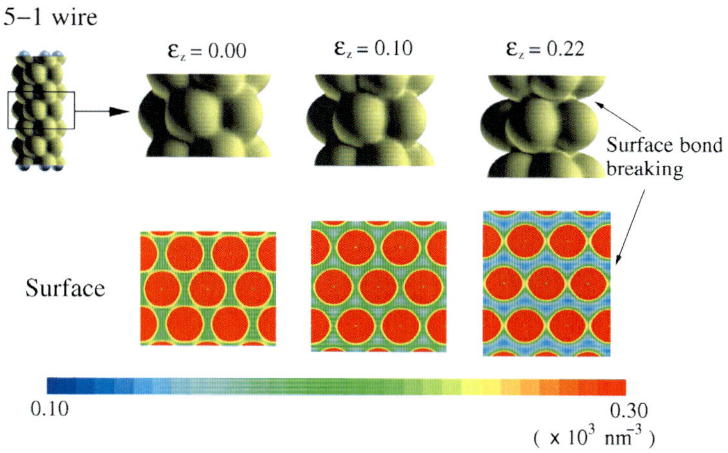

Figure 7.21 Change in atomic and electronic structures of a Cu multi-shell 5-1 nanowire during tensile loading (the same calculation as in Fig. 7.18). Reprinted from Kushima, A., et al. (2006) "Ideal strength of a Cu multi-shell nano-wire," *Mod. Simul. Mater. Sci. Eng.*, 14, pp. 1031–1039, with permission from IOP Publishing.

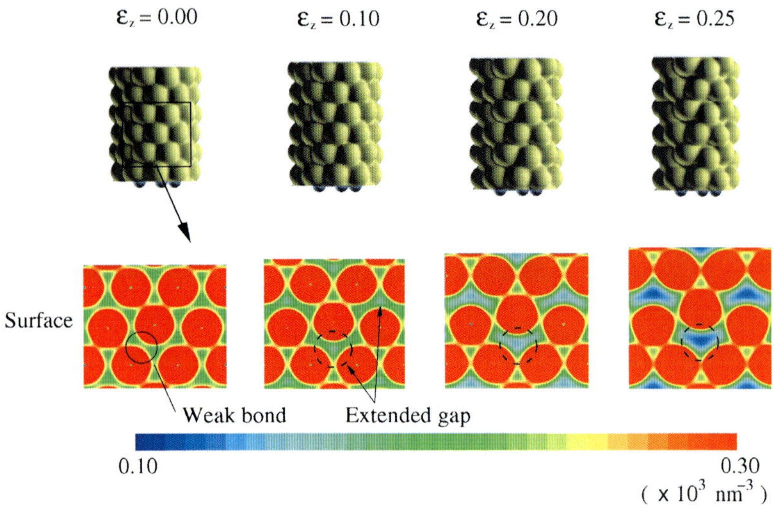

Figure 7.22 Change in atomic and electronic structures of a Cu multi-shell 10-5-1 nanowire during tensile loading (the same calculation as in Fig. 7.18). Reprinted from Kushima, A., et al. (2006) "Ideal strength of a Cu multi-shell nano-wire," *Mod. Simul. Mater. Sci. Eng.*, 14, pp. 1031–1039, with permission from IOP Publishing.

7.7 MULTI-PHYSICS PROPERTIES OF NANO-ELEMENTS

The electric property is never constant when materials are subject to external load or constraint but changes—sometimes dramatically—because of the rearrangement in the atomic structure and electronic state. This coupling effect between mechanical deformation and electric properties is called the multi-physics property. Strained silicon is a well-known example of the multi-physics property. Applied strain changes the band structure of silicon, and at the critical strain silicon undergoes a semiconducting-to-metallic transition [Umeno and Kitamura, 2002]. The insulator-metallic transition can be observed in carbon nanotubes [Umeno et al.,

2004a; Umeno et al., 2004b] and SiC polytypes [Umeno et al., 2007] as well. Not only is there a change of bandgap, but various electrical properties, e.g., ferroelectricity and magnetism, change as well. As mentioned in the preceding section, nano-elements that have a peculiar atomic configuration such as nanowires and atom-chains, the microstructure (the atomic structure of which surface and crystal structure) are quite different from their bulk counterparts in this regard. Hence, nano-elements often exhibit characteristic multi-physics properties that cannot be achieved in a bulk. In this section, we focus on the multi-physics properties of nano-structures and review typical electronic properties: bandgap and ferroelectricity.

7.6.3 Metallic-Insulator Transition in Semiconductors

There is an energy bandgap between the valence band and the conduction band in a semiconductor such as silicon. When this energy bandgap is large, electrons cannot move and electric current does not flow, and the semiconductor behaves as an insulator. In contrast, when the energy bandgap becomes small (or disappears), electrons move from the valence band to the conduction band, suggesting that the material behaves as a conductor. First-principles calculations based on the density functional theory (DFT) can evaluate the band structure of materials, allowing a simulation to follow the change of the bandgap as a strain is applied to the nano-structure. However, as is well known, the DFT underestimates the energy bandgap to a lower than an experimentally observed magnitude, although it is still worth discussing with regard to qualitative behavior. In recent years, a theory that overcomes the problem of underestimation, e.g., self-interaction correction (SIC) [Perdew and Zunger, 1981, Svane and Gunnarsson 1988] and GW approximation [Hedin, 1965], has been proposed, and it is likely that the quantitative discussion will continue in the future.

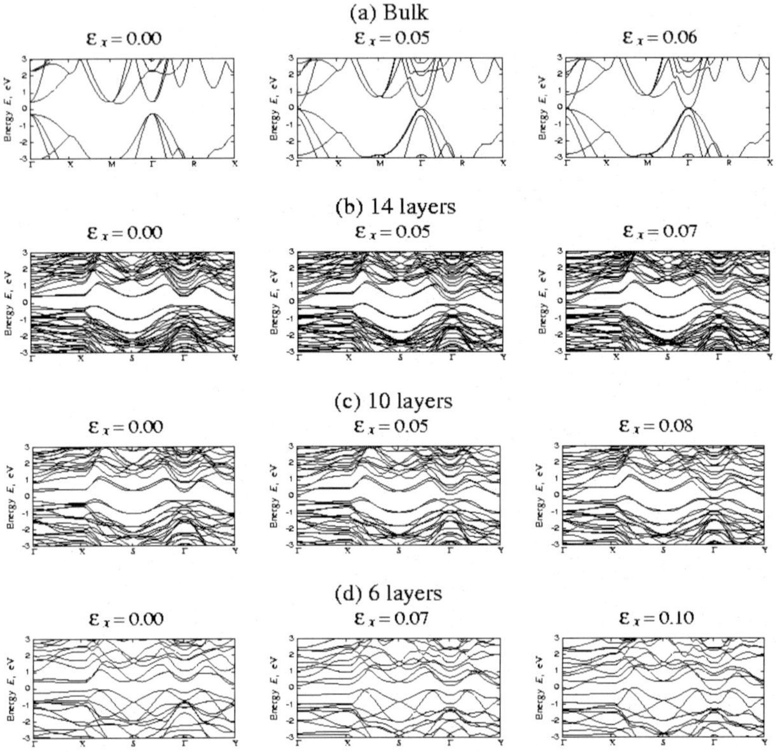

Figure 7.23 Change in the electronic band structures of (a) Si bulk and (b) 14-layer, (c) 10-layer, and (d) 6-layer Si slabs under [011] tension (*ab initio* calculations with the ultrasoft pseudopotential within the GGA). Reprinted with permission from Umeno, Y., et al. (2005) "*Ab initio* study of the surface properties and ideal strength of (100) silicon thin films," *Phys. Rev. B*, 72, article number 165431. Copyright (2005) by the American Physical Society.

Figure 7.23 shows the band structure of Si bulk and film consisting of 14, 10, or 6 atomic layers with a (100) surface in response to tensile strain along the [011] direction [Umeno et al., 2005], where the simulation models are the same as in Fig. 7.9. The pseudopotentials based on Vanderbilt's ultrasoft method within the generalized gradient approximation (GGA) are employed for the first-principles calculations. The wave number vector of the abscissa axis indicates the symmetric points in the Brillouin zone.

In as unstrained bulk material, the maximum energy of the valence band is situated around the Gamma point and the minimum of energy of the conduction band is around the M point. In a thin film, on the other hand, both of these spread around the fermi energy, and the maximum energy of the valence band is around the X point, while the minimum energy of the conduction band is around the X' point. When tension is applied, the declination of the conduction band at the Gamma point and the raise of the valence band cause the energy bandgap energy in the bulk material to decrease. In contrast, there is no remarkable change under tension in nano-films. Figure 7.24 shows the energy bandgap, E_{gap}, in a Si bulk material and in Si nano-film in response to tensile strain. Although the Si bulk initially has a large energy bandgap, the gap decreases as tensile strain is applied, and it finally disappears at around a strain of ε = 0.06. This indicates that silicon undergoes a semiconductor-conductor transition. Although the Si nano-film has an initially smaller energy bandgap than that of bulk, the decrease in the energy bandgap under tensile strain is small. Thus, nano-films keep their energy bandgap at a relatively large strain.

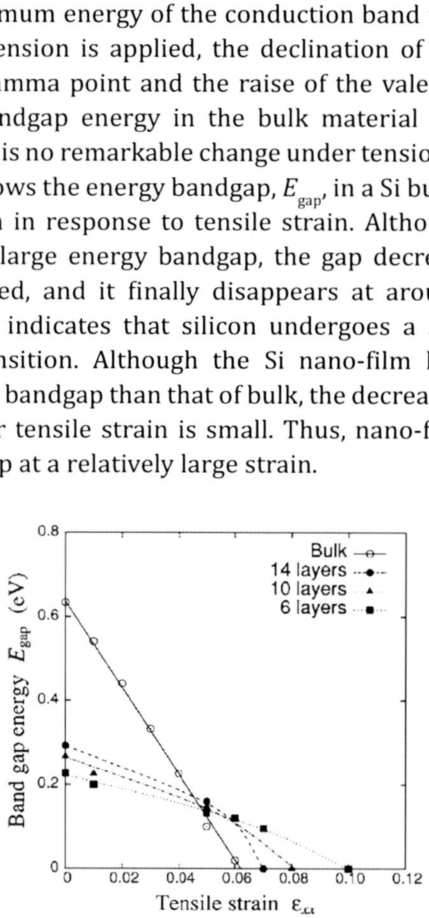

Figure 7.24 Energy bandgap of Si bulk and 14-, 10-, and 6-layer slabs as a function of [011] tensile strain (the same calculation as in Fig. 7.23). Reprinted with permission from Umeno, Y., et al. (2005) "*Ab initio* study of the surface properties and ideal strength of (100) silicon thin films," *Phys. Rev. B*, 72, article number 165431. Copyright (2005) by the American Physical Society.

One of the typical nano-structures that exhibit the conducting-insulating transition is the carbon nanotube (CNT) [Yang et al., 1999]. CNTs are cylindrically rolled-up graphene sheets. CNTs display the characteristic property of a semiconductor or conductor, depending on which direction the graphite sheets are rolled, namely the arrangement direction (chirality) of a carbon six-membered ring. Figure 7.25 shows the atomic structure and the energy bandgap of a zigzag (9, 0) CNT as a function of the axial tensile strain [Kushima, 2007]. The pseudopotentials based on the USPP method and the generalized gradient approximation (GGA) are employed for the *ab initio* calculation. At the equilibrium state, there is about a 0.15 eV energy bandgap, indicating the CNT is initially a semiconductor. As tensile strain is applied, the E_{gap} increases, but it begins to decrease from $\varepsilon = 0.15$ and finally disappears at $\varepsilon = 0.25$. This indicates that the CNT has become a conductor. For other CNTs, a transition not only from semiconductor to conductor but also from conductor to semiconductor has been found. In addition, the transition occurs under compressive strain as well as tensile strain along the circumferential direction of the CNT [Umeno et al., 2004a]. This conducting-insulating transition has been evaluated in experiments [Bai et al., 2007] and applied to electronic devices such as micro wiring with CNTs.

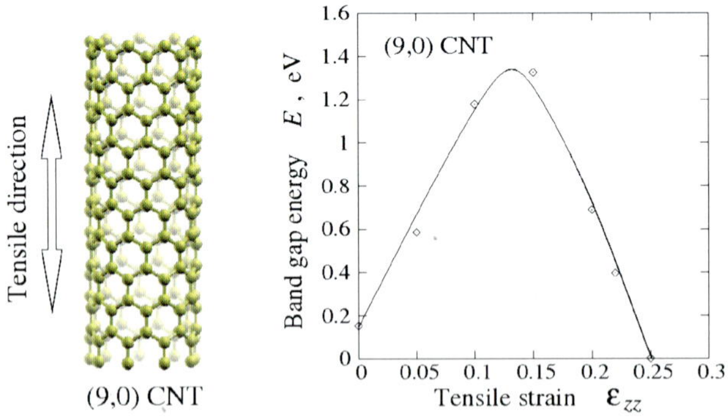

Figure 7.25 Energy bandgap of (9, 0) CNT as a function of axial tensile strain (*ab initio* calculations with the ultrasoft pseudopotential within the GGA). Reprinted with permission from Kushima, A. (2007) "*Ab initio* studies on ideal strength of low-dimensional structures," PhD Thesis, Kyoto University, pp. 39–46.

7.6.4 Multi-physics Properties in Ferroelectric Materials

Ferroelectric perovskite-type oxides such as lead titanate, $PbTiO_3$, exhibit spontaneous polarization (which is defined as a dipole moment per unit volume) because the octahedral consisting of oxygen atoms is buckled by the delicate balance between long-range Coulomb interactions and the short-range covalent (see Fig. 7.26). Ferroelectric materials have drawn much attention because of their technological applications, e.g., non-volatile random access memory (FeRAM) and piezoelectric devices.

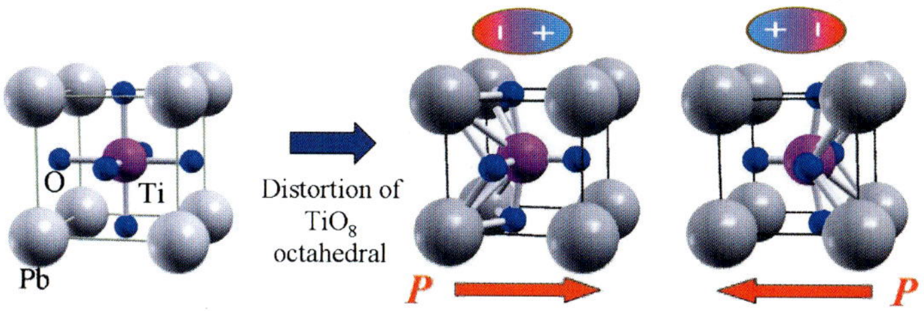

Figure 7.26 The atomic structure of ferroelectric perovskite $PbTiO_3$.

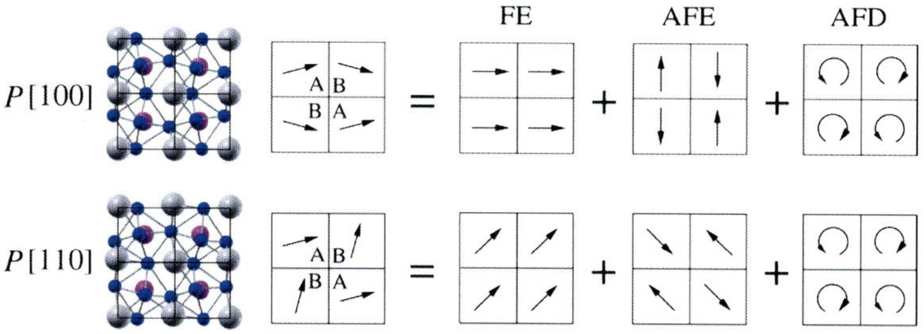

Figure 7.27 The atomic structure of c(2×2) reconstructed (001) surfaces of the PbO termination. The polarization direction lies along the [100] (top) and [110] (bottom). Arrows indicate the direction of atomic displacement in the ferroelectric (FE), antiferroelectric (AFE), and antiferrodistortive (AFD) phases (*ab initio* calculations with the PAW pseudopotentials within the LDA). Reprinted with permission from Umeno, Y., et al. (2006) "*Ab initio* density functional theory study of strain effects on ferroelectricity at $PbTiO_3$ surfaces," *Phys. Rev. B*, 74, article number 174111. Copyright (2006) by the American Physical Society.

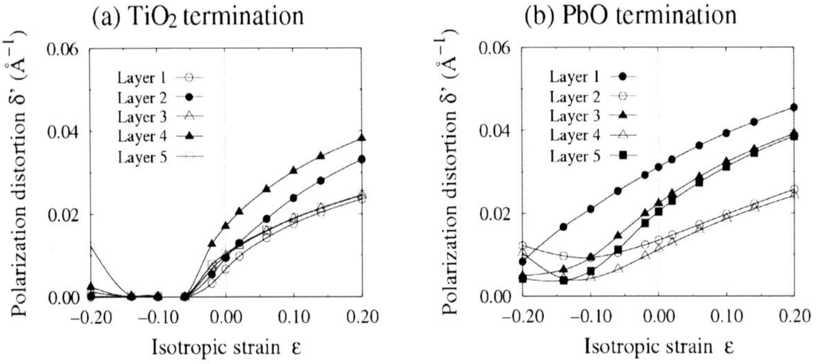

Figure 7.28 Polarization distortion of 5 atomic layers at TiO_2 and PbO terminated surfaces without surface reconstruction as a function of in-plane isotropic strain (the same calculation as Fig. 7.27). Reprinted with permission from Umeno, Y., et al. (2006) "*Ab initio* density functional theory study of strain effects on ferroelectricity at $PbTiO_3$ surfaces," *Phys. Rev. B*, 74, article number 174111. Copyright (2006) by the American Physical Society.

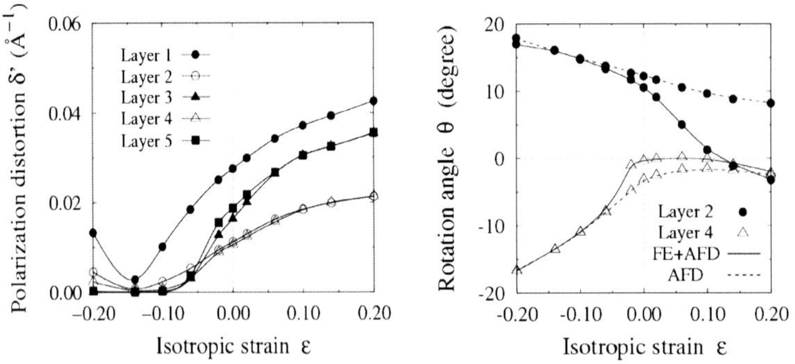

Figure 7.29 Polarization distortion and AFD rotation angle of 5 atomic layers at the reconstructed PbO-terminated surfaces as a function of in-plane isotropic strain (the same calculation as Fig. 7.27). Reprinted with permission from Umeno, Y., et al. (2006) "*Ab initio* density functional theory study of strain effects on ferroelectricity at $PbTiO_3$ surfaces," *Phys. Rev. B*, 74, article number 174111. Copyright (2006) by the American Physical Society.

Since the coordination number is quite low at the surface, each atom shifts in a direction perpendicular to the surface. This brings about a bumpy surface called a rumpling structure. In addition, these peculiar atom shifts cause complex surface reconstruction in $PbTiO_3$ called antiferroelectric (AFE) and antiferrodistortive (AFD) materials (Fig. 7.27) [Umeno et al., 2006]. This surface reconstruction disarranges spontaneous polarization at the surface, and ferroelectric properties absolutely differ from those of bulk.

Figure 7.28 shows the ferroelectric properties of a $PbTiO_3$ nano-film consisting of nine atomic layers with (001) surface in response to in-plane isotropic tensile or compressive strain [Umeno et al., 2006]. It shows the results of two possible surface terminations, one of the PbO atomic layer and one of the TiO_2 atomic layer. The pseudopotentials based on the projector-augmented wave (PAW) method [Blochl, 1994] and the local density approximation (LDA) method [Ceperley and Alder, 1980] are used for the *ab initio* calculations. At equilibrium, the ferroelectricity is suppressed in the TiO_2-terminated surface. On the other hand, the ferroelectricity is enhanced in the PbO-terminated surface, which indicates remarkable dependence on the surface termination. Under the strained state, polarization (ferroelectricity) increases in response to the tensile in-plane strain, while polarization decreases in response to compression. However, in a TiO_2-terminated surface, the polarization in each atomic layer is suppressed, and it disappears at the critical strain of $\varepsilon = -0.05$. In contrast, in the PbO-terminated surface, the polarization is enhanced and remains even under high compression.

The left side of Fig. 7.29 shows the change in polarization in a reconstructed PbO-terminated ferroelectric nano-film in response to in-plane isotropic tensile or compressive strain. As in a nonreconstructed PbO-terminated surface (Fig. 7.19), the ferroelectricity increases in response to tensile strain and decreases in response to compressive strain. However, the polarization is entirely lower than that in the nonreconstructed surface, and the trend is much more remarkable in response to compressive strain.

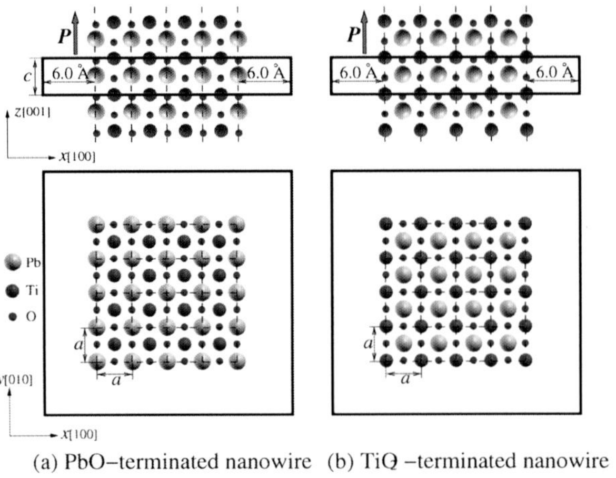

Figure 7.30 Simulation models of the (a) PbO-terminated and (b) TiO_2-terminated nanowires with a cross-section of 4×4 unit cells. Reprinted with permission from Shimada, T., et al. (2009) "*Ab initio* study of ferroelectricity in edged $PbTiO_3$ nanowires under axial tension," *Phys. Rev. B*, 79, article number 024102. Copyright (2009) by the American Physical Society.

The right side of Fig. 7.29 shows the change in rotation angle of a TiO_8 octahedral by the reconstruction, the AFD. It appears remarkably in response to compressive strain. This suggests that ferroelectricity is suppressed under compressive strain, which in turn indicates the importance of AFD surface reconstruction. Thus, ferroelectricity is affected strongly by the internal structure (surface terminations and reconstructions) with the external structure (surface).

$PbTiO_3$ nanowires, as one-dimensional nanoscale ferroelectric material [Shimada et al., 2009], have been manufactured with a diameter of a few nanometers. The important feature of nanowire is on the influences of surface and surface/surface interaction. In addition, nanowires have an atomically sharp edge structure consisting of (100) and (010) surfaces. In particular, since the coordination number is much lower at the edge compared with the bulk material, the atomic configuration and electronic state change characteristically, which strongly affect the ferroelectricity. Figure 7.30 shows $PbTiO_3$ nanowires with these edge structures. On the

$PbTiO_3$ surface, there are two possible terminations of the PbO and TiO_2 atomic layers. Each nanowire has a cross-section in which four perovskite unit cells, denoted as the cross-section of 4 × 4 unit cells below, are arranged in the *x* and *y* directions. The thin nanowires with 3 × 3, 2 × 2, and 1 × 1 models are also studied. Pseudopotentials based on the projector-augmented wave (PAW) method and the local density approximation (LDA) are used for *ab initio* calculations. Table 7.3 lists the lateral lattice parameter along the *x*, *y* direction; the axial lattice parameter along the *z* direction, *a* and *c*; and the averaged polarization, *P*, after relaxation of the atomic structure. As the size of the PbO-terminated nanowire decreases, the averaged polarization increases, and even thinnest PbO-terminated nanowire with a 1 × 1 unit cell size exhibits spontaneous polarization. In addition, the smallest nanowire has the highest magnitude. In contrast, the polarization of the TiO_2-terminated nanowire is highly suppressed. The nanowires thinner than 4 × 4 unit cell in cross-section do not exhibit spontaneous polarization, indicating the TiO_2-terminated edge strongly suppresses the ferroelectric distortions. Thus, there exists clear size-dependence and critical size for the electronic property.

Table 7.3 Averaged lateral and axial lattice parameters, *a* and *c* (in angstroms), and averaged polarization *P* (in μ C/cm^2) in both the PbO-terminated and TiO_2-terminated nanowires as a function of cross-sectional size [Shimada et al., 2009]

Nanowire size	1×1	2×2	3×3	4×4	Bulk
		PbO termination			
$\bar{a}$	3.60	3.64	3.70	3.73	3.87
c	3.92	3.97	3.99	4.01	4.03
$c/\bar{a}$	1.088	1.090	1.081	1.074	1.043
$\bar{P}$	103.7	99.9	95.7	94.3	85.8
		TiO_2 termination			
$\bar{a}$	3.77	3.81	3.84	3.85	3.87
c	3.39	3.57	3.67	3.74	4.03
$c/\bar{a}$	0.900	0.935	0.956	0.973	1.043
$\bar{P}$	0.00	0.00	0.00	0.89	85.8

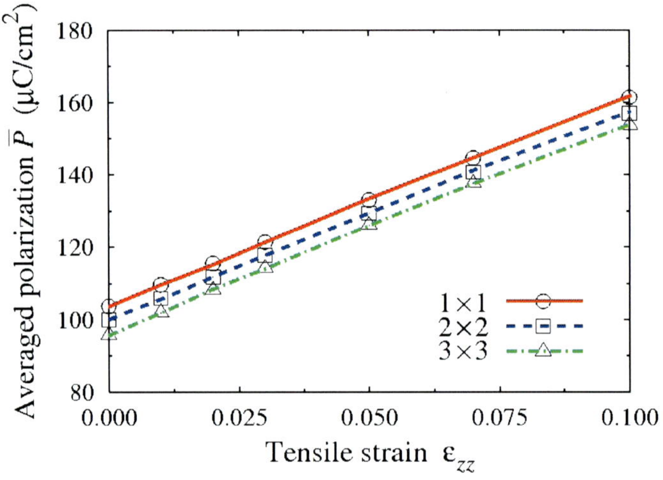

(a) PbO-terminated nanowires

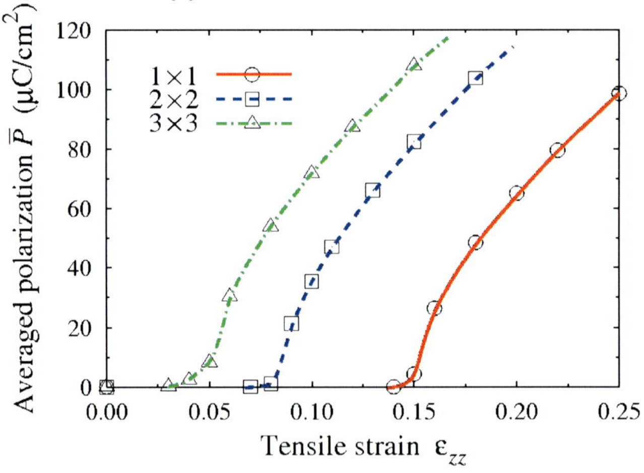

(b) TiO_2-terminated nanowires

Figure 7.31 Average polarization of the PbO-terminated and TiO_2-terminated nanowires as a function of axial tensile strain (*ab initio* calculations with the PAW pseudopotential within LDA). Reprinted with permission from Shimada, T., et al. (2009) "*Ab initio* study of ferroelectricity in edged $PbTiO_3$ nanowires under axial tension," *Phys. Rev. B*, 79, article number 024102. Copyright (2009) by the American Physical Society.

Figure 7.31 plots the averaged polarization, *P*, in the each nanowire as a function of the axial tensile strain. Averaged polarization increases in response to the tensile strain for all the PbO-terminated nanowires. For the TiO_2-terminated nanowires which are initially paraelectric, the ferroelectricity appears under axial tension. Thinner nanowires require larger critical strain.

Using the multi-physics property, we can control various electric properties in nano-structures by controlling the mechanical stress or strain.

References

Andersen, H. C. (1980) Molecular dynamics simulations at constant pressure and/or temperature, *J. Chem. Phys.*, 72, pp. 2384–2393.

Bahr, D. F., Kramer, D. E., and Gerberich, W. W. (1998) Non-linear deformation mechanisms during nanoindentation, *Acta Mater.*, 46, pp. 3605–3617.

Bai, X., Golberg, D., Bando, Y., Zhi, C., Tang, C., Mitome, M., and Kurashima, K. (2007) Deformation-driven electrical transport of boron nitride nanotubes, *Nano Lett.*, 7, pp. 632–637.

Baskes, M. I. (1992) Modified embedded-atom potentials for cubic materials and impurities, *Phys. Rev. B*, 46, pp. 2727–2742.

Blochl, P. E. (1994) Projector augmented-wave method, *Phys. Rev. B*, 50, pp. 17953–17979.

Brenner, D. W. (1990) Empirical potential for hydrocarbons for use in simulating the chemical vapor deposition of diamond films, *Phys. Rev. B*, 42, pp. 9458–9471.

Ceperley, D. M., and Alder, B. J. (1980) Ground state of the electron gas by a stochastic method, *Phys. Rev. Lett.*, 45, pp. 566–569.

Cerny, M., Sob, M., Pokluda, J., and Sandera, P. (2004) *Ab initio* calculations of ideal tensile strength and mechanical stability in copper, *J. Phys.: Cond. Mat.*, 16, pp. 1045–1052.

Clatterbuck, D. M., Chrzan, D. C., and Morris, J. W. (2001) The inherent tensile strength of iron, *Philos. Mag. Lett.*, 82, article number 135501.

Clatterbuck, D. M., Krenn, C. R., Cohen, M. L., and Morris, J. W. (2003) Phonon instabilities and the ideal strength of aluminum, *Phys. Rev. Lett.*, 91, article number 135501.

Daw, M. S., and Baskes, M. I. (1984) Embedded-atom method: derivation and application to impurities, surfaces, and other defects in metals, *Phys. Rev. B*, 29, pp. 6443–6453.

Dick, B. J., and Overhauser, A. W. (1958) Theory of the dielectric constants of alkali halide crystals, *Phys. Rev.*, 112, pp. 90–103.

Esposito, E., Carlson, A. E., Ling, D. D., Ehrenreich, H., and Gelatt, Jr., C. D. (1980) First-principles calculations of the theoretical tensile strength of copper, *Philos. Mag. A*, 41, pp. 251–259.

Finnis, M. W., and Sinclair, J. E. (1984) A simple N-body potential for transition metals, *Philos. Mag. A*, 54, pp. 45–55.

Friák, M., Sob, M., and Vitek, V. (2003) *Ab initio* calculation of tensile strength in iron, *Philos. Mag.*, 83, pp. 3529–3537.

Gear, C. W. (1971) *Numerical Initial Value Problems in Ordinary Differential Equations*, Prentice Hall, Englewood Cliffs, NJ.

Gouldstone, A., Koh, H. J., Zeng, K. Y., Giannakopoulos, A. E., and Suresh, S. (2000) Discrete and continuous deformation during nanoindentation of thin films, *Acta Mater.*, 48, pp. 2277–2295.

Greer, J. R. (2006) Bridging the gap between computational and experimental length scales: a review on nanoscale plasticity, *Rev. Adv. Mater. Sci.*, 13, pp. 59–70.

Hedin, L. (1965) New method for calculating the one-particle Green's function with application to the electron-gas problem, *Phys. Rev.*, 139, pp. A796–A823.

Hohenberg, P., and Kohn, W. (1964) Inhomogeneous electron gas, *Phys. Rev.*, 136, pp. B864–B871.

Huang, Y. M., Spence, J. C. H., and Sankey, O. F. (1994) The effect of impurities on the ideal tensile strength of silicon, *Philos. Mag.*, 70, pp. 53–62.

Kelly, A., and Macmillan, N. H. (1986) *Strong Solids*, Clarendon Press, Oxford.

Kitamura, T., and Umeno, Y. (2003) Validity of effective medium theory for aluminium under tension, *Model. Simul. Mat. Sci. Eng.*, 11, pp. 127–136.

Kitamura, T., Umeno, Y., and Tsuji, N. (2003) Internal atomic stress near sigma5 tilt grain boundary in aluminum under tension, *Mod. Simul. Mater. Sci. Eng.*, 11, pp. 839–849.

Kohn, W., and Sham, L. (1965) Self-consistent equations including exchange and correlation effects, *Phys. Rev.*, 140, pp. A1133–A1138.

Kondo, Y., and Takayanagi, K. (2000) Synthesis and characterization of helical multi-shell gold nanowires, *Science*, 289, pp. 606–608.

Kushima, A. (2007) *Ab initio* studies on ideal strength of low-dimensional structures, PhD Thesis, Kyoto University, pp. 39–46.

Kushima, A., Umeno, Y., and Kitamura, T. (2006) Ideal strength of a Cu multi-shell nano-wire, *Mod. Simul. Mater. Sci. Eng.*, 14, pp. 1031–1039.

Li, W., and Wang, T. (1998) *Ab initio* investigation of the elasticity and stability of aluminium, *J. Phys Condens. Matter,* 10, pp. 9889–9904.

Luo, W. D., Roundy, D., Cohen, M. L., and Morris, J. W. (2002) Ideal strength of bcc molybdenum and niobium, *Phys. Rev. B*, 66, article number 094110.

Macmillan, N. H. (1983) The ideal strength of solids, in *Atomistics of Fracture*, ed. Latanision, R. M., and Pickens, J. R., Plenum Press, New York, p. 95.

Milstein, F. (1973) Theoretical strength of a perfect crystal with exponentially attractive and repulsive interatomic interaction, *J. Appl. Phys.*, 44, pp. 3833–3840.

Morris, Jr., Krenn, J. W., Roundy, D., and Cohen, M. L. (2000) Ideal shear strengths of fcc aluminum and copper, *Proceedings of the Hume-Rothery Symposium, in honor of A. G. Khatchatuyan*, TMS, Warrendale, Penn., pp. 187–207.

Nakatani, K., Nakatani, A., and Kitagawa, H. (1998) Molecular dynamics study on fracture mechanism of Fe-amorphous metal (*J* integral near mode I crack tip), *Adv. Mat. Res.*, 1, pp. 88–98.

Nose, S. (1984) A molecular dynamics method for simulations in the canonical ensemble, *Mol. Phys.*, 52, pp. 255–268.

Ogata, S., and Kitagawa, H. (1999) *Ab initio* tensile testing simulation of aluminum and aluminum nitride ceramics based on density functional theory, *Comp. Mater. Sci.*, 15, pp. 435–440.

Ohnishi, H., Kondo, Y., and Takayanagi, K. (1998) Quantized conductance through individual rows of suspended gold atoms, *Nature*, 395, pp. 780–783.

Parrinello, M., and Rahman, A. (1981) Polymorphic transitions in single crystals: a new molecular dynamics method, *J. Appl. Phys.*, 52, pp. 7182–7190.

Perdew, J. P., and Wang, Y. (1992) Accurate and simple analytic representation of the electron-gas correlation energy, *Phys. Rev. B*, 45, pp. 13244–13249.

Perdew, J. P., and Zunger, A. (1981) Self-interaction correction to density-functional approximations for many-electron systems, *Phys. Rev. B*, 23, pp. 5048–5079.

Ray, J. R., and Rahman, A. (1984) Statistical ensembles and molecular dynamics studies of anisotropic solids, *J. Chem. Phys.*, 80, pp. 4423–4428.

Roundy, D., and Cohen, M. L. (2001) Ideal strength of diamond, Si, and Ge, *Phys. Rev. B*, 64, article number 212103.

Roundy, D., Krenn, C. R., Cohen, M. L., and Morris, J. W. (2001) The ideal strength of tungsten, *Philos. Mag. A*, 81, pp. 1725–1747.

Sandera, P., and Pokluda, J. (1994), Ideal strength of cubic-crystals under triaxial tension, *J. Metall. Mat.*, 32, pp. 180–189.

Sandera, P., and Pokluda, Improvement of the Mackenzie theory on ideal shear-strength, J. (1993) *Scr. Metall. Mat.*, 29, pp. 1445–1450.

Shan, Z. W., Adesso, G., Cabot, A., Sherburne, M. P., Asif, S. A. S., Warren, O. L., Chrzan, D. C., Minori, A. M., and Alivisatos, A. P. (2008) Ultrahigh stress and strain in hierarchically structured hollow nanoparticles, *Nat. Mater.*, 7, pp. 947–952.

Shimada, T., Tomoda, S., and Kitamura, T. (2009) *Ab initio* study of ferroelectricity in edged $PbTiO_3$ nanowires under axial tension, *Phys. Rev. B*, 79, article number 024102.

Shimada, T., Wakahara, K., Umeno, Y., and Kitamura, T. (2008) Shell model potential for $PbTiO_3$ and its applicability to surfaces and domain walls, *J. Phys. Condens. Matter*, 20, article number 325225.

Sob, M., Wang, L. G., and Vitek, V. (1998), Ab initio calculation of the ideal tensile strength in copper and nickel aluminide, *Metall. Mat.*, 36, pp. 145- 151.

Sob, M., Wang, L. G., and Vitek, V. (1997) Theoretical tensile stress in tungsten single crystals by full-potential first-principles calculations, *Mater. Sci. Eng. A*, 234–236, pp. 1075–1078.

Stillinger F. H., and Weber, T. A. (1985) Computer simulations of local order in condensed phases of silicon, *Phys. Rev. B*, 31, pp. 5262–5271.

Svane, A., and Gunnarsson, O. (1988) Localization in the self-interaction-corrected density-functional formalism, *Phys. Rev. B*, 37, pp. 9919–9922.

Telling, R. H., Pickard, C. J., Payne, M. C., and Field, J. E. (2000) Theoretical strength and cleavage of diamond, *Phys. Rev. Lett.*, 84, pp. 5160–5163.

Tersoff, J. (1988a) New empirical approach for the structure and energy of covalent systems, *Phys. Rev. B*, 37, pp. 6991–7000.

Tersoff, J. (1988b) Empirical interatomic potential for silicon with improved elastic properties, *Phys. Rev. B*, 38, pp. 9902–9905.

Umeno, Y., and Kitamura, T. (2002) *Ab initio* simulation on ideal shear strength of silicon, *Mater. Sci. Eng. B*, 88, pp. 79–84.

Umeno, Y., Kinoshita, Y., and Kitamura, T. (2007) *Ab initio* DFT simulation of ideal shear deformation of SiC polytypes, *Mod. Simul. Mater. Sci. Eng.*, 15, pp. 27–37.

Umeno, Y., Kitamura, T., and Kushima, A. (2004a) Theoretical analysis on electronic properties of zigzag-type single-walled carbon nanotubes under radial deformation, *Comp. Mater. Sci.*, 30, pp. 283–287.

Umeno, Y., Kitamura, T., and Kushima, A. (2004b) Metallic-semiconducting transition of single-walled carbon nanotubes under high axial strain, *Comp. Mater. Sci.*, 31, pp. 33–41.

Umeno, Y., Kushima, A., Kitamura, T., Gumbsch, P., and Li, J. (2005) *Ab initio* study of the surface properties and ideal strength of (100) silicon thin films, *Phys. Rev. B*, 72, article number 165431.

Umeno, Y., Shimada, T., Kitamura, T., and Elsaesser, C. (2006) *Ab initio* density functional theory study of strain effects on ferroelectricity at $PbTiO_3$ surfaces, *Phys. Rev. B*, 74, article number 174111.

Vanderbilt, D. (1990) Soft self-consistent pseudopotentials in a generalized eigenvalue formalism, *Phys. Rev. B*, 41, pp. 7892–7895.

Verlet, L. (1967) Computer "experiments" on classical fluids. I. thermodynamical properties of Lennard-Jones molecules, *Phys. Rev.*, 159, pp. 98–103.

Wallace, D. C. (1972) *Thermodynamics of Crystals*, Wiley, New York.

Wang, C. Z., Chen, C. T., and Ho, K. M. (1989) Empirical tight-binding force model for molecular-dynamics simulation of Si, *Phys. Rev. B*, 39, pp. 8586-8592.

Woodcock, C. L., and Bahr, D. F. (2000) Plastic zone evolution around small-scale indentations, *Scripta Mater.*, 43, pp. 783–788.

Yang, L., Anantram, M. P., Han, J., and Lu, J. P. (1999) Band-gap change of carbon nanotubes: effect of small uniaxial and torsional strain, *Phys. Rev. B*, 60, pp. 13874–13878.

Yu, M. F., Lourie, O., Dyer, M. J., Moloni, K., Kelly, T. F., and Ruoff, R. S. (2000) Strength and breaking mechanism of multiwalled carbon nanotubes under tensile load, *Science*, 287, pp. 637–640.

Zhou, W. X., Johnson, R. A., and Wadley, H. N. G. (2004) Misfit-energy-increasing dislocations in vapor-deposited CoFe/NiFe multilayers, *Phys. Rev. B*, 69, article number 144113.

Chapter 8

FRACTURE MECHANICS IN ATOMIC COMPONENTS

8.1 STRESS SINGULARITY AT THE ATOMIC SCALE

The size of a singular stress region near an interface edge is dependent on the film thickness [Becker et al., 1997; Kitamura et al., 2003]. The thinner the film, the smaller the stress-concentrated region becomes. In a component of an advanced device, where the thickness is in the order of several tens of nanometers or less, the region size must be on a scale of several nanometers. In such a case, the applicability of the analysis based on the continuum theory is questionable because of the discreteness originating from atomic structure (the atomic distance is about 0.2–0.3 nm).

Atomic simulations such as molecular dynamics simulations, described in the previous chapter, are powerful tools for investigating the fracture process and mechanics in detail, taking into account the influence of discreteness. Many atomic simulations have been carried out attempting to analyze interface strength [e.g., Baskes et al., 1994; Gall et al., 2000; Suga et al., 1991; Kubo et al., 1999] or crack propagation [e.g., Paskin et al., 1980; Mullins and Dokainish, 1982; Celis et al., 1983; Cheung and Yip, 1990]. However, most of these simulations targeted the microscopic fracture process (e.g., emission of dislocation) of a macro-scale component; they are multi-scale simulations. They set

Fracture Nanomechanics
Edited by Takayuki Kitamura, Hiroyuki Hirakata, Takashi Sumigawa and Takahiro Shimada

www.panstanford.com

a boundary condition of stress/strain based on continuum mechanics, and the applicability to nano-scale materials was out of their scope. Then, little has been discussed about the local stress (singularity) field or the validity of the fracture criterion for nano-scale or atomic-scale components.

In this section, we investigate using molecular dynamics in the stress field and the fracture criterion of atomic-scale components.

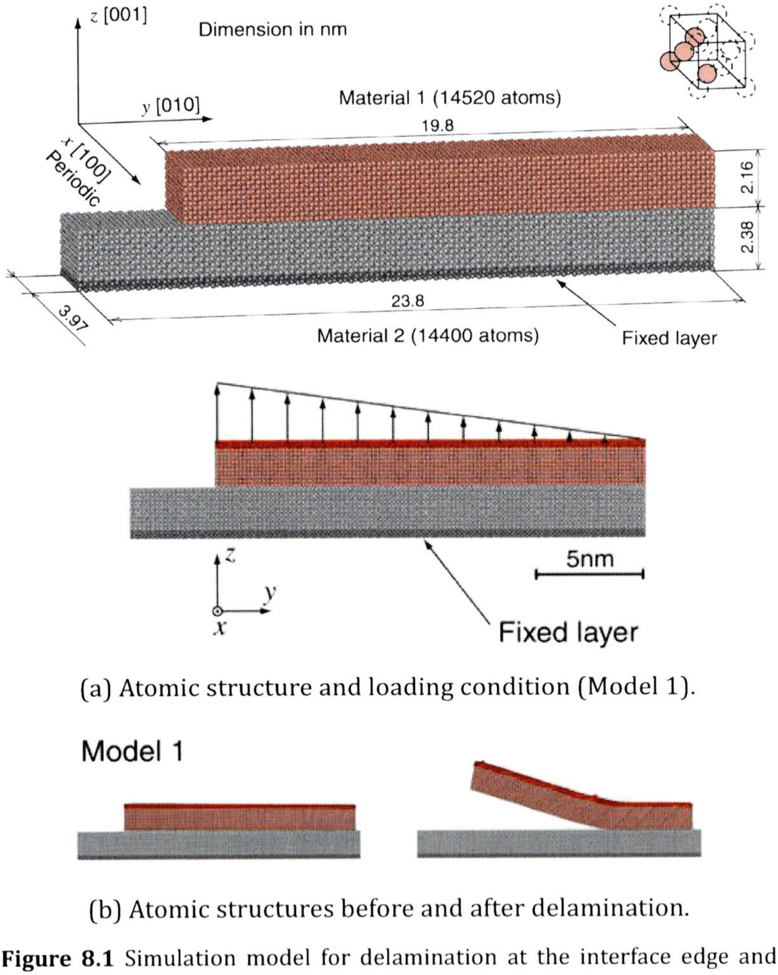

Figure 8.1 Simulation model for delamination at the interface edge and applied load. Reprinted with permission from Kitamura, T., et al. (2004) "Applicability of fracture mechanics on brittle delamination of nano-scale film edge," *JSME Int. J. Ser. A*, 47, pp. 106–112.

Figure 8.1a shows a simulation of an interface delamination crack initiating from an interface edge calculated by the atomic simulation (Model 1) [Kitamura et al., 2004a]. The dissimilar nanomaterial is composed of copper thin film with a thickness of about 2 nm on a substrate of tungsten (W), where the bottom of the W layer is fixed, and the periodic boundary condition is applied in the x direction. Here, the Morse potential is employed to describe the interaction between atoms because the elastic properties are well reproduced despite its simple functional form, and the local stress is easily evaluated in the atomic resolution [Girifalco and Weizer, 1959], as described in Section 7.4.2.1. Obviously, continuum mechanics points to the existence of stress singularity along the interface near the interface edge. The stress can be numerically analyzed using the FEM, where the minimum element size of mesh must be at an atomic scale or smaller. This follows the mathematical assumption for the examination of continuum mechanics concept even though it is not physical. As shown in Fig. 8.1b, a crack initiates from the interface edge and propagates along the interface without dislocation; i.e., brittle fracture.

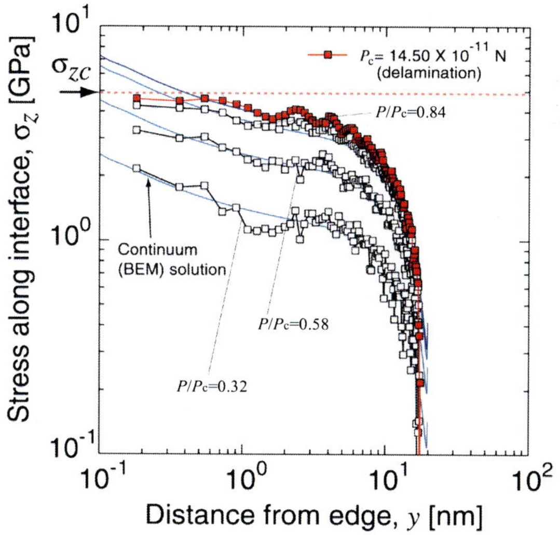

Figure 8.2 Stress distribution along the interface under loading (Model 1). Reprinted with permission from Kitamura, T., et al. (2004a) "Applicability of fracture mechanics on brittle delamination of nano-scale film edge," *JSME Int. J. Ser. A*, 47, pp. 106–112.

Figure 8.2 shows the normal stress distribution along the interface at several load levels. The broken lines, which indicate the singular field, show the stress distribution under the particular loads calculated by the FEM analysis assuming linear-elastic deformation of both materials. At a lower applied load (P/P_c = 0.32; P_c is the critical load at the delamination crack initiation), the atomic stress distribution coincides well with the FEM solution, although there is a slight fluctuation in the atomic stress. At a middle load (P/P_c = 0.58), a similar coincidence is found except very near the interface edge (r < 0.5 nm). In other words, the stress begins to differ from that expected in continuum mechanics in the high stress region at this load level. At a high load (P/P_c = 0.84), there is fairly good coincidence, taking into account the fluctuation, while the discrepancy zone expands near the edge. At the critical load (P/P_c = 1.00), stress in r < 1 nm differs from that expected in continuum mechanics and is almost constant. Since we can assume that this is the fracture process zone, we can still apply conventional fracture mechanics concept to this delamination considering the scale of component (about 20 nm).

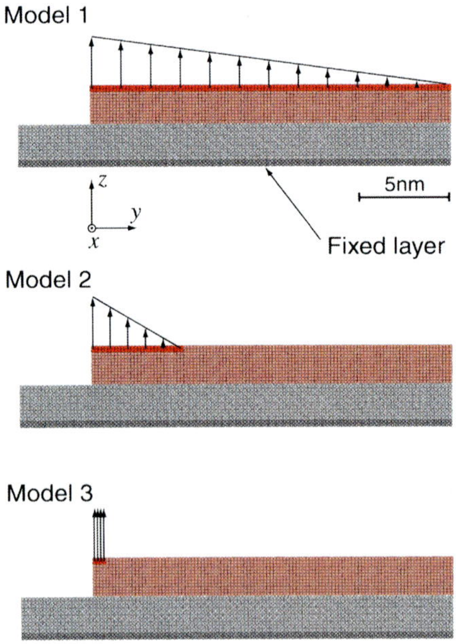

(a)

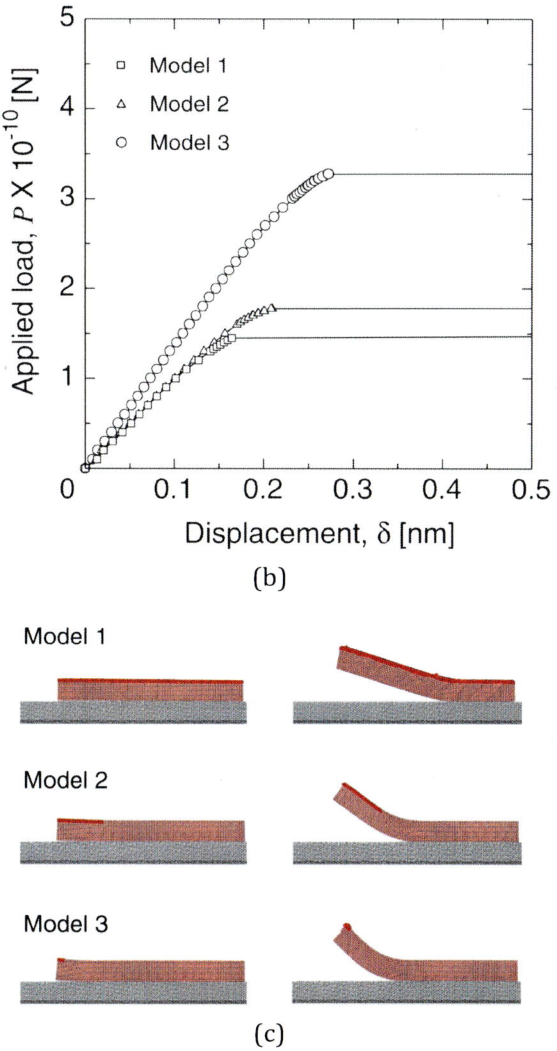

Figure 8.3 (a) Loading condition, (b) load-displacement curves, and (c) atomic structures before and after delamination. Reprinted with permission from Kitamura, T., et al. (2004a) "Applicability of fracture mechanics on brittle delamination of nano-scale film edge," *JSME Int. J. Ser. A*, 47, pp. 106–112.

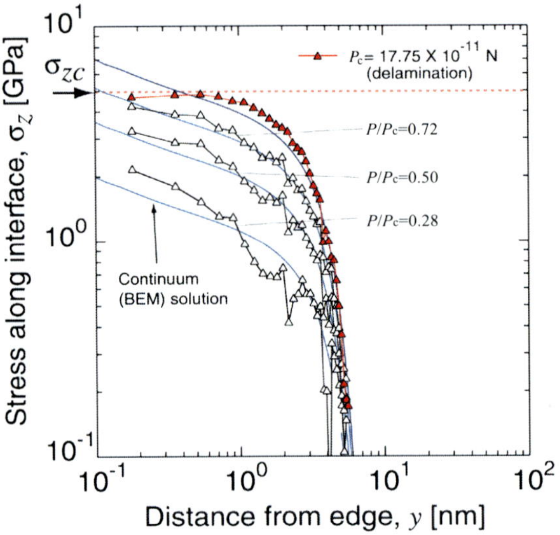

(a) Model 2

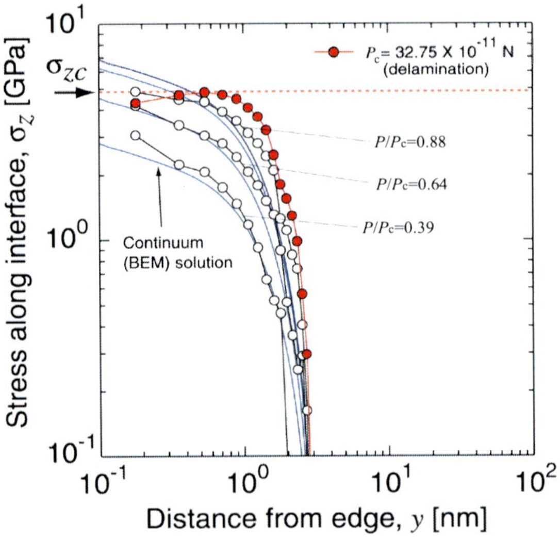

(b) Model 3

Figure 8.4 Stress distribution along the interface under the loading for (a) Model 2 and (b) Model 3. Reprinted with permission from Kitamura, T., et al. (2004a) "Applicability of fracture mechanics on brittle delamination of nano-scale film edge," *JSME Int. J. Ser. A*, 47, pp. 106–112.

Figure 8.3 shows the load-displacement curves in the delamination simulations of different models (Fig. 8.3a: Models 1, 2, and 3) under several applied loads. The δ increases almost proportionally with *P*, while a slight nonlinearity originating from the interatomic potential at the high-strain condition is observed before the rapid increase in δ. If the nonlinearity is properly included in the FEM analysis shown in Fig. 8.2, it might give us a better match in the stress field between atomic mechanics and continuum mechanics at the critical load in the vicinity of the edge. Figure 8.4 shows the normal stress distribution along the interface at several loading levels of Models 2 and 3 (see also Fig. 8.2a). In spite of the difference in loading modes and the resulting P-δ curves, the atomic stress distribution obtained by the atomic simulation coincides fairly well with the stress distribution given by the FEM solution based on continuum mechanics even in high-load conditions ($0.72 < P/P_c < 0.88$).

Figure 8.5 shows the critical stress distributions ($P/P_c = 1.00$) under the three types of loading. The stress converges in $r < 1$ nm in all the models. This suggests more clearly how well fracture mechanics applies in nano-components. In the figure, σ_{zc} stands for the tensile strength of the interface in a bulk material, which is obtained by an atomic simulation for a model with a three-dimensional periodic boundary condition of the same combination of the dissimilar materials. Thus, the delamination takes place when the stress in $r < 1$ nm (about six atoms from the interface edge) reaches the ideal strength of the interface, σ_{zc}. These indicate that delamination is governed by the stress field of the six atoms (the region of 1 nm) and the criterion is given by the ideal strength σ_{zc} in this region.

In the analysis, weak interface strength is assumed so that brittle fracture takes place. If we assume a strong interface strength, we may have dislocation emission before delamination. However, the critical condition at the dislocation emission and the applicability of fracture mechanics might be discussed in a similar way.

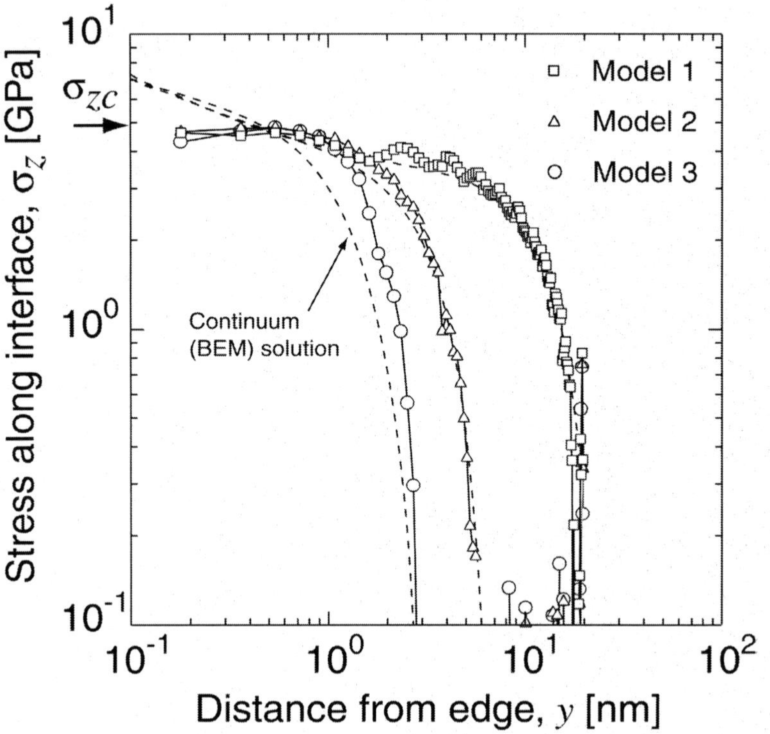

Figure 8.5 Stress distribution along the interface at delamination for Models 1, 2, and 3. Reprinted with permission from Kitamura, T., et al. (2004a) "Applicability of fracture mechanics on brittle delamination of nano-scale film edge," *JSME Int. J. Ser. A*, 47, pp. 106–112.

8.2 INSTABILITY CRITERION FOR AN ATOMIC STRUCTURE

Material strength is fundamentally understood in terms of bond breaking and/or switching at the atomic level of crystals. For example, brittle fracture of cleavage of the crack tip is clearly associated with bond breaking, and the characteristics of multiple dislocations govern plastic deformation.

Bond switching is the mechanical instability of the atomic lattice structure where the material deforms under a sudden drop of the external load. In recent years, small atomic components such as nano-films, nano-tubes (wires), and nano-dots have been developed thanks to remarkable advancements in manufacturing technology. Single bond switching in a component sometimes causes the fatal malfunction of a device; therefore, it is important to understand the instability characteristics of atomic structures.

In this section, we show the instability criterion of atomic structures starting with the Born stability criteria for perfect crystals and concluding with the rigorous instability criterion for arbitrary (inhomogeneous) atomic components.

8.2.1 Instability Criterion of Perfect Crystal Lattice

8.2.1.1 Lattice instability based on elastic stiffness tensors (*B*-criterion)

According to Born [Born and Huang, 1954], the stability of a perfect crystal is determined by the positivity of the determinant of its elastic constant matrix $\{C_{ij}\}$. For a cubic crystal, the stability criterion is given as

$$\begin{aligned} &C_{11} + 2C_{12} > 0, \\ &C_{44} > 0, \\ &C_{11} - C_{12} > 0 \end{aligned} \tag{8.1}$$

where C_{ij} denotes the elastic constant in the Voigt notation. The first and second criteria correspond to the bulk modulus, $B_T = (C_{11} + 2C_{12})/3$, and shear modulus, $G = C_{44}$, being positive, respectively. The third criterion is comprehended as the positive modulus in the tetragonal shear, $G' = (C_{11} - C_{12})/2$. These are referred as the Born stability criteria and are used to predict the melting point [Hunter, 1942], polymorphism [Milstein and Farber, 1980], and pressure-induced amorphization [Tse and Klug, 1991]. Since the criterion is derived for a perfect crystal in a stress-free condition, it is not adequate for investigating stability under a strain.

Extending the Born stability criteria, an instability criterion of a perfect crystal under finite strain has been proposed using an elastic stiffness coefficient [Wang et al., 1993]. Let us consider an applied uniform stress, taking a crystal in an arbitrary initial configuration **X** to a final configuration ***x***. The corresponding Lagrangian strain is $\eta_{ij} = (u_{ij} + u_{ji} + u_{ij}u_{ji})/2$ where $u_{ij} = du_i/dX_j$ and $\boldsymbol{u} = \boldsymbol{x} - \mathbf{X}$. The elastic stiffness coefficient, $B_{ijkl}(\mathbf{X})$, which is defined as $B_{ijkl}(\mathbf{X}) \equiv [\partial\sigma_{ij}(\mathbf{X})/\partial\eta_{kl}]_{\mathbf{X}}$ where $\sigma_{ij}(\mathbf{X})$ is the applied stress, is written as

$$B_{ijkl}(\mathbf{X}) = C_{ijkl}(\mathbf{X}) + \frac{1}{2}[\sigma_{il}(\mathbf{X})\delta_{jk} + \sigma_{jl}(\mathbf{X})\delta_{ik} + \sigma_{ik}(\mathbf{X})\delta_{jl} + \sigma_{jk}(\mathbf{X})\delta_{il} - 2\sigma_{ij}(\mathbf{X})\delta_{kl}] \quad (8.2)$$

where $C_{ijkl}(\mathbf{X})$ denotes the elastic constants evaluated at the applied stress. The stability under a homogeneous deformation is evaluated from the positivity of the determinant of elastic stiffness coefficient matrix, $\{B_{ijkl}(\mathbf{X})\}$ (B-criterion). Note that the elastic stiffness coefficient does not have Voigt symmetry. Thus, it is recommended that we use

$$\mathbf{B}' = \frac{1}{2}(\mathbf{B}^T + \mathbf{B}) \quad (8.3)$$

For a perfect cubic crystal under hydrostatic stress, $P = \sigma_{11} = \sigma_{22} = \sigma_{33}$, the elements of **B** are $B_{11} = B_{22} = B_{33} = C_{11} - P$, $B_{12} = B_{21} = B_{13} = C_{12} + P$, and $B_{44} = B_{55} = B_{66} = C_{44} - P$. The stability criterion is described as

$$\begin{aligned} M_1 &= (C_{11} + 2C_{12})/3 + P/3 > 0 \\ M_2 &= C_{44} - P > 0 \\ M_3 &= (C_{11} - C_{12})/2 - P > 0 \end{aligned} \quad (8.4)$$

M_1, M_2, and M_3, are referred as spinodal instability (in the sense of vanishing of B_T), shear instability, and Born instability, respectively. Figure 8.6 compares Born stability and B-criterion applied to gold bulk using the embedded-atom method (EAM) potential under increasing hydrostatic pressure at a constant temperature of T = 500 K. The simulation cell contains 504 Au atoms arranged in an fcc structure under a three-dimensional periodic boundary condition. According to the Born stability criteria (the open circle and triangles), the system becomes unstable at $(a/a_0)_{th} = 1.025$ via the vanishing of tetragonal shear, where a_0 denotes the initial lattice parameter in a stress-free condition. However, the actual instability observed by direct simulation takes place at $(a/a_0)_{obs} = 1.053$, while the system is still stable at $(a/a_0) = 1.025$. On the other hand, the *B*-criterion (the solid circles and triangles) points out the critical strain of $(a/a_0)_{th} = 1.059$ via the vanishing of bulk modulus, which agrees well with $(a/a_0)_{obs} = 1.053$.

As mentioned above, the *B*-criterion is valid for investigating mechanical instability in a perfect crystal under a finite strain. However, it cannot describe the stability of atomic components with an inhomogeneous structure, e.g., a body with a crack and a defect.

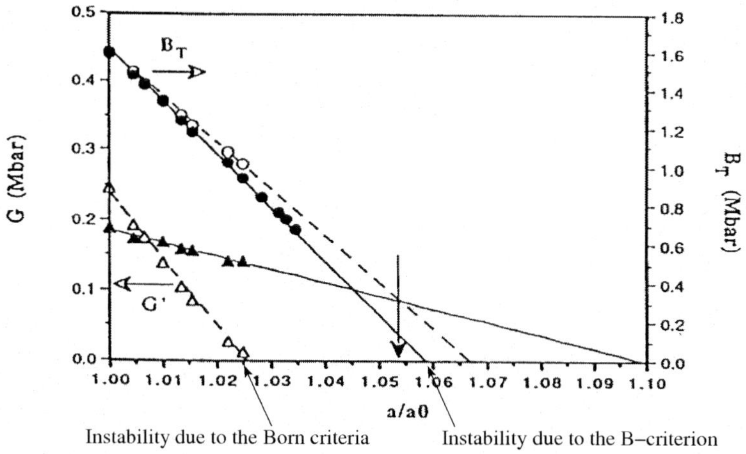

Figure 8.6 Variations of M_1 (closed circles) and M_3 (closed triangles), as defined by the *B*-criterion in Eq. (8.4), with lattice strain a/a_0. Solid lines indicate linear extrapolations to give critical values for each instability. Also shown are B_T (open circles) and G' (open triangles) from the Born stability criteria in Eq. (8.1). Arrow indicates the critical strain observed in direct simulations. Reprinted with permission from Wang, J., et al. (1993) "Crystal instabilities at finite strain," *Phys. Rev. Lett.*, 71, pp. 4182–4185. Copyright (1993) by the American Physical Society.

8.2.1.2 Soft phonon mode criterion (*P*-criterion)

A phonon is a quantized mode of atomic (lattice) vibration occurring in the three-dimensional crystal lattice of a solid. The phonon dispersion relation is a fundamental materials property in lattice dynamics. The phonon vibration frequency and its eigenmode are obtained by solving the eigenvalue problem of the following dynamical matrix, $D(\mathbf{k})$:

$$D(\mathbf{k}) = \sum_{m} \frac{1}{\sqrt{M_\alpha M_\beta}} K_{ni\alpha}^{mj\beta} \exp[i\mathbf{k} \,.\, (\mathbf{R}_{n\beta} - \mathbf{R}_{m\alpha})] \tag{8.5}$$

where the subscripts m and n denote numbers of unit cells, α and β are the indexes of an atom in a unit cell, i and j are the indexes in the x, y, z direction, $\mathbf{R}_{m\alpha}$ is the coordinate of the α-th atom in the m-th cell, M_α is the mass of the α-th atom, and $\mathbf{k}$ is the wave number vector. The interatomic force constant, $K_{ni\alpha}{}^{mj\beta}$, is defined as the second-order derivative of the potential energy of the system, Φ, with respect to the atomic coordinate,

$$K_{ni\alpha}^{mj\beta} = \frac{\partial^2 \Phi}{\partial \mathrm{r}_{ni\alpha} \partial \mathrm{r}_{mj\beta}} \tag{8.6}$$

Since the dynamical matrix, $D(\mathbf{k})$, in Eq. (8.5) is a function of the wave number vector, $\mathbf{k}$, the phonon eigenfrequency and the eigenmode depends on the $\mathbf{k}$. Hence, the phonon dispersion relation ($\mathbf{k} - \omega$) is obtained by diagonalizing the dynamical matrix, $D(\mathbf{k})$, changing the wave number vector, $\mathbf{k}$, within the Brillouin zone. Here, the reciprocal lattice vectors of the real-space primitive translation vectors representing a unit cell, $\mathbf{a}_1$, $\mathbf{a}_2$, and $\mathbf{a}_3$, for the Brillouin zone are

$$\mathbf{b}_1 = 2\pi \frac{\mathbf{a}_2 \times \mathbf{a}_3}{\mathbf{a}_1 \cdot (\mathbf{a}_2 \times \mathbf{a}_3)}$$

$$\mathbf{b}_2 = 2\pi \frac{\mathbf{a}_3 \times \mathbf{a}_1}{\mathbf{a}_1 \cdot (\mathbf{a}_2 \times \mathbf{a}_3)}$$

$$\mathbf{b}_3 = 2\pi \frac{\mathbf{a}_1 \times \mathbf{a}_2}{\mathbf{a}_1 \cdot (\mathbf{a}_2 \times \mathbf{a}_3)} \tag{8.7}$$

If the eigenfrequency, ω, is a real number, the atoms stably vibrate with the corresponding eigenmode. This is because the potential energy surface along the eigenmode displacement has a convex downward curvature (see Fig. 8.7a). On the other hand, if the eigenfrequency, ω, is an imaginary number, namely, the potential curvature is convex upward, the atomic displacement gradually diverges along the corresponding eigenmode direction and finally the original atomic structure collapses (see Fig. 8.7b). Thus, the imaginary eigenfrequency gives the instability criterion of an atomic structure. The criterion is called phonon soft mode or *P*-criterion.

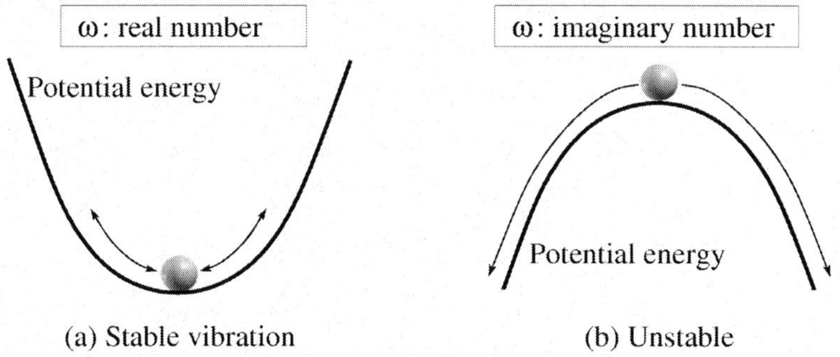

Figure 8.7 Schematic illustration explaining the soft phonon mode. (a) Stable phonon vibration (the eigenfrequency ω is real number). (b) Soft phonon mode (the eigenfrequency ω is imaginary number).

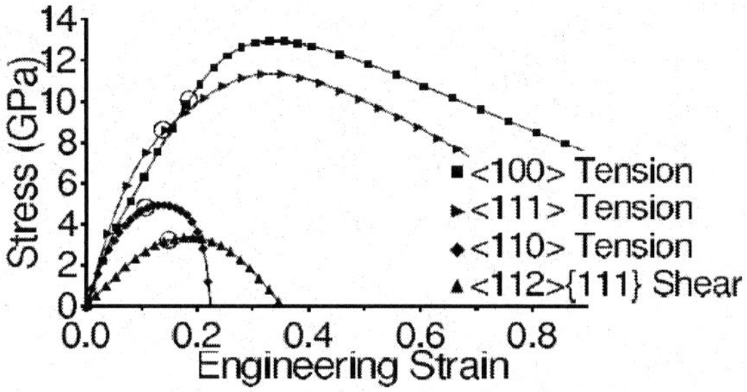

Figure 8.8 Stress-strain curves of Al perfect fcc crystal lattice. Open circles indicate the critical strain, which is evaluated as unstable by soft phonon mode analysis. Reprinted with permission from Clatterbuck, D. M., et al. (2003) "Phonon instabilities and the ideal strength of aluminum," *Phys. Rev Lett.*, 91, article number 135501. Copyright (2003) by the American Physical Society.

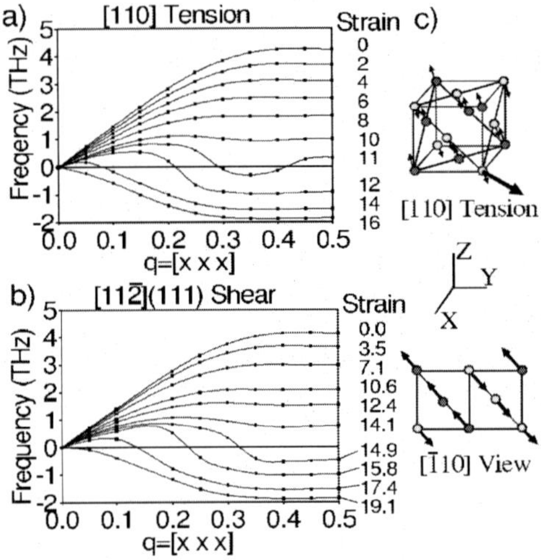

Figure 8.9 Phonon frequencies as a function of strain for (a) [110] tension and (b) [11$\bar{2}$] (111) shear. (c) A schematic illustration of the instability mode. Arrows indicate unstable soft mode displacement. In both cases the crystallography of the instability is the same.Reprinted with permission from Clatterbuck, D. M., et al. (2003) "Phonon instabilities and the ideal strength of aluminum," *Phys. Rev Lett.*, 91, article number 135501. Copyright (2003) by the American Physical Society.

Table 8.1 Stress and strain associated with phonon instabilities as compared to the peak in the stress-strain curves shown in Fig. 8.7 [Clatterbuck et al., 2003]. Also shown are the shear plane and displacement associated with phonon instabilities.

Loading configuration	Phonon instability σ (GPa)	e	Peak stress-strain curve σ (GPa)	e	Shear system of instability
[110] tension	4.89	0.115	4.92	0.140	[$\bar{1}$12](111)
[11$\bar{2}$](111) shear	3.16	0.145	3.33	0.185	[$\bar{1}$12](111)
[001] tension	9.20	0.170	12.92	0.340	[001](100)
					[1$\bar{1}$0](11$\bar{1}$)
[111] tension	8.95	0.150	11.30	0.330	[111](001)
					[111](110)
					[111]($\bar{1}$10)
					[111]($\bar{1}$11)

The phonon soft mode criterion has been used to analyze the bifurcations of crystals under finite deformation and pressure-induced phase transformation. Figure 8.8 shows the stress-strain curves of a perfect Al crystal under homogeneous tension or shear obtained from first-principles calculations with the norm-conserving pseudopotentials within the local density approximation (LDA) [Clatterbuck et al., 2003]. At each strain, the phonon frequency is calculated on the basis of the density-functional perturbation theory (DFPT). The open circles in the figure show the unstable critical strain evaluated from the phonon soft mode criterion. The result suggests that the perfect Al crystal becomes mechanically unstable before the stress reaches its maximum (ideal strength). Figure 8.9a,b shows the change in the phonon eigenfrequency under [110] tension and [112] (111) shear, respectively. The eigenfrequency decreases as the applied strain increases, and it finally reaches an imaginary number at a specific wave number vector. Figure 8.9c depicts the eigenmode displacement corresponding to the soft phonon mode. The soft mode under the [110] tension shows that shear-type instability along the <112> direction on the (111) plane occurs. Table 8.1 summarizes the critical stress for the soft phonon mode, the peak stress, and the instability mode from the phonon soft mode under each deformation path. The phonon soft mode instability appears before reaching peak stress. Thus, the criterion is an effective way to investigate the instability of the internal atomic displacement as well, something which cannot be found by the *B*-criterion.

Although the phonon assumes that the perfect crystal has three-dimensional periodicity, Dmitriev et al. have successfully extended the concept to two-dimensional (nano-films) and one-dimensional (nano-fibers) systems to investigate instability in low-dimensional nano-components [Dmitriev et al., 2005].

8.2.3 Instability Criterion in Inhomogeneous Atomic Structure

Some attempts have been made to evaluate the stability of inhomogeneous atomic structures. For example, the dislocation nucleation in a Ni nano-wire under tension has been examined by applying the B-criterion to a local crystal lattice [Yashiro and Tomita, 2001]. There has also been an investigation on the soft phonon mode criterion for a local site, which is known as the Λ-criterion [Van Vliet et al., 2003]. Van Vliet et al. discussed dislocation nucleation in a nano-indentation simulation based on this criterion. Although the conditions of instability have been estimated to a certain extent through the criteria do not maintain rigorousness because only the energy balance in a local region is taken into account. To precisely evaluate mechanical instability in an arbitrary atomic system, it is necessary to consider all the degrees of freedom of the entire atomic system.

In this section, we describe an analytical method that considers the energy balance with respect to all the degrees of freedom of atoms.

8.2.2.1 Rigorous expression of instability criterion in an atomic system

A universal and rigorous criterion has been proposed for investigating mechanical instability in an arbitrary atomic structure under any loading/constraint condition [Kitamura et al., 2004b]. Unlike the B-criterion where there are only limited degrees of freedom (9 degrees of freedom for the primitive translation vectors), the instability criterion considers all the degrees of freedom that a system intrinsically has.

Let us consider an atomic system consisting of N atoms. The number of degrees of freedom, M, is $3N - 6$, excluding rigid-body translation and rotation. An arbitrary deformation of the system can be described by the $3N - 6$ dimensional vector, $\mathbf{R}$, consisting of atomic coordinate

$$\begin{aligned}\mathbf{R} &\equiv^t (r_z^{\,2}, r_y^{\,3}, r_z^{\,3}, r_x^{\,4}, r_y^{\,4}, r_z^{\,4},\ldots, r_i^{\alpha},\ldots, r_x^{\,N}, r_y^{\,N}, r_z^{\,N}) \\ &=^t (R_1, R_2,\ldots, R_m,\ldots, R_M) \qquad (8.8)\end{aligned}$$

where r_i^{α} is the coordinate of atom α in the i direction. The potential energy of the system can be represented as a function of the vector, $\mathbf{R}$,

$$\Phi = \Phi(\mathbf{R}) \tag{8.9}$$

Note that the total number of degrees of freedom is $M = 3N + 3$ when the system is subject to a three-dimensional periodic boundary condition.

Let us consider the atomic equilibrium system at 0 K under a constant external load. Then, the total energy of the system, Π, is the sum of the potential energy, Φ, and the work done by the external load, W ($\Pi = \Phi + W$). Taylor's series expansion of Π in terms of infinitesimal deformation, $\Delta\mathbf{R}$, is in the form

$$\begin{aligned}\Pi(\mathbf{R} + \Delta\mathbf{R}) &= \Pi(\mathbf{R}) + \sum_{m=1}^{M} \frac{\partial \Phi}{\partial R_m} \Delta R_m + \sum_{m=1}^{M} \frac{\partial W}{\partial R_m} \Delta R_m \\ &+ \frac{1}{2} \sum_{m=1}^{M} \sum_{n=1}^{M} \frac{\partial^2 \Phi}{\partial R_m \partial R_n} \Delta R_m \Delta R_n + \ldots \\ &+ \frac{1}{2} \sum_{m=1}^{M} \sum_{n=1}^{M} \frac{\partial^2 W}{\partial R_m \partial R_n} \Delta R_m \Delta R_n + \ldots \end{aligned} \tag{8.10}$$

The second and third terms on the right-hand side are zero because the system is at equilibrium and the external load is constant. Ignoring the higher-order terms, the change in the total energy against the infinitesimal deformation, $\Delta\Pi$, can be given by

$$\Delta\Pi = \Pi(\mathbf{R} + \Delta\mathbf{R}) - \Pi(\mathbf{R}) = \frac{1}{2} \sum_{m=1}^{M} \sum_{n=1}^{M} \frac{\partial^2 \Phi}{\partial R_m \partial R_n} \Delta R_m \Delta R_n \tag{8.11}$$

If $\Delta\Pi$ is always positive against any infinitesimal deformation, $\Delta\mathbf{R}$, the atomic system is stable. On the other hand, if $\Delta\Pi$ is negative against a $\Delta\mathbf{R}$, the system becomes unstable and the corresponding deformation progresses consequently without an increase in external load. Here, we introduce a Hessian matrix, $\mathbf{A}$, consisting of second-order derivatives with respect to the atomic coordinate,

$$A_{mn} = \frac{\partial^2 \Phi}{\partial R_m \partial R_n} \tag{8.12}$$

Eq. (8.10) is rewritten as

$$\Delta\Pi = \frac{1}{2}\Delta\mathbf{R}^T \mathbf{A}\Delta\mathbf{R} \tag{8.13}$$

Now, thinking about the eigenvalue problem of the Hessian matrix, **A**,

$$\mathbf{A}\mathbf{p}_m = \eta_m \mathbf{p}_m \quad (m = 1, 2, \ldots, M) \tag{8.14}$$

we obtain the M eigenvalues, η_m ($\eta_1 < \eta_2 < \cdots < \eta_M$), and M eigenvectors, $\boldsymbol{p}_m$. Since the eigenvectors are a complete orthogonal basis set of the system, the arbitrary infinitesimal deformation, $\Delta\mathbf{R}$, can be represented using the eigenvectors as

$$\Delta\mathbf{R} = \sum_{m=1}^{M} \Delta Q_m \mathbf{p}_m \tag{8.15}$$

Using Eqs. (8.12) and (8.14), $\Delta\Pi$ is rewritten as

$$\Delta\Pi = \frac{1}{2}\sum_{m=1}^{M} \eta_m (\Delta Q_m)^2 \tag{8.16}$$

In this equation, the eigenvalue, η_m, signifies the potential energy curvature against the deformation along the corresponding eigenvector, $\boldsymbol{p}_m$. Therefore, if the minimum eigenvalue is positive ($\eta_1 > 0$), the change in the total energy, $\Delta\Pi$, is always positive with respect to any infinitesimal deformation, indicating that the system is mechanically stable. On the other hand, when the minimum eigenvalue becomes negative ($\eta_1 \leq 0$), the total energy can decrease along the corresponding eigenvector, $\boldsymbol{p}_1$, indicating the atomic motion at the beginning of instability (unstable mode).

The validity of the method has been confirmed by applying it to various inhomogeneous atomic structures such as a crack [Kitamura et al., 2004b], interface [Kitamura et al., 2004c], and amorphous metals [Umeno et al., 2007].

Figure 8.11 shows the simulation model of a cracked body with the external load controlled (Case A) and with the displacement controlled (Case B). Figure 8.10a plots the stress-strain curves of Cases A and B. The two curves show the different behavior (especially, the different critical strain). Note that these bring about different morphology of mechanical instability after the critical strain: the cleavage crack propagation in Case A (brittle fracture) and the dislocation emission from the crack tip in Case B (ductile fracture). The instability criterion, $\eta_1 = 0$, exactly corresponds to each critical strain where the stress drops (or the strain diverges). These prove the validity of the criterion.

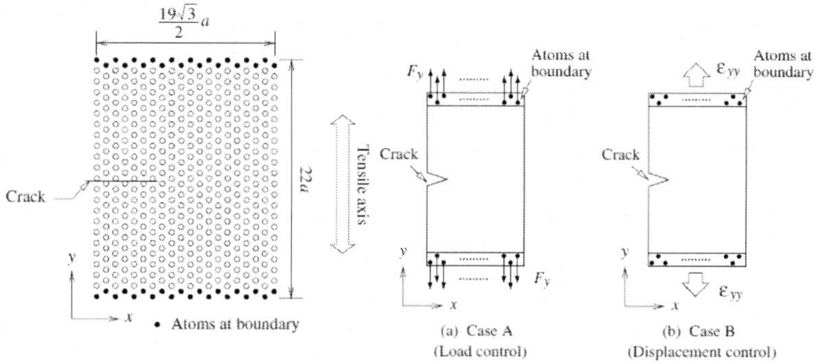

Figure 8.10 Two-dimensional simulation model with a crack. (a) Case A: load control simulation. (b) Case B: displacement control simulation. Reprinted from Kitamura, T., et al. (2004) "Analytical evaluation of unstable deformation criterion of atomic structure and its application to nanostructure," *Comp. Mater. Sci.*, 29, pp. 499–510, Copyright (2004), with permission from Elsevier.

Figure 8.12 shows the unstable deformation mode corresponding the minimum eigenvalue of η_1=0. The dominant mode vectors in the Cases A and B represent the cleavage of crack and the dislocation emission from the crack tip, respectively, which agrees with the results directly obtained in the molecular dynamics simulations. Thus, the method can rigorously evaluate the instability criterion as well as the deformation mode of the mechanical instability.

The weak point of this method is the difficulty of applying it to a large-scale atomic structure because the Hessian matrix, **A**, solved in the method becomes too large to compute its eigenvalues.

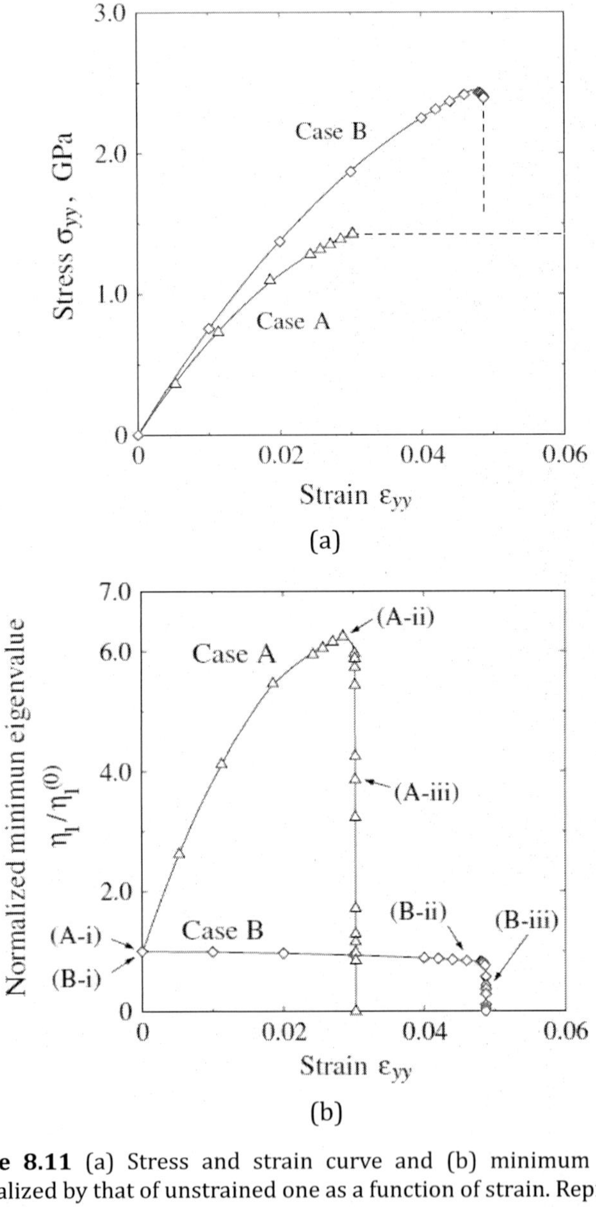

Figure 8.11 (a) Stress and strain curve and (b) minimum eigenvalue normalized by that of unstrained one as a function of strain. Reprinted from Kitamura, T., et al. (2004) "Analytical evaluation of unstable deformation criterion of atomic structure and its application to nanostructure," *Comp. Mater. Sci.*, 29, pp. 499–510, Copyright (2004), with permission from Elsevier.

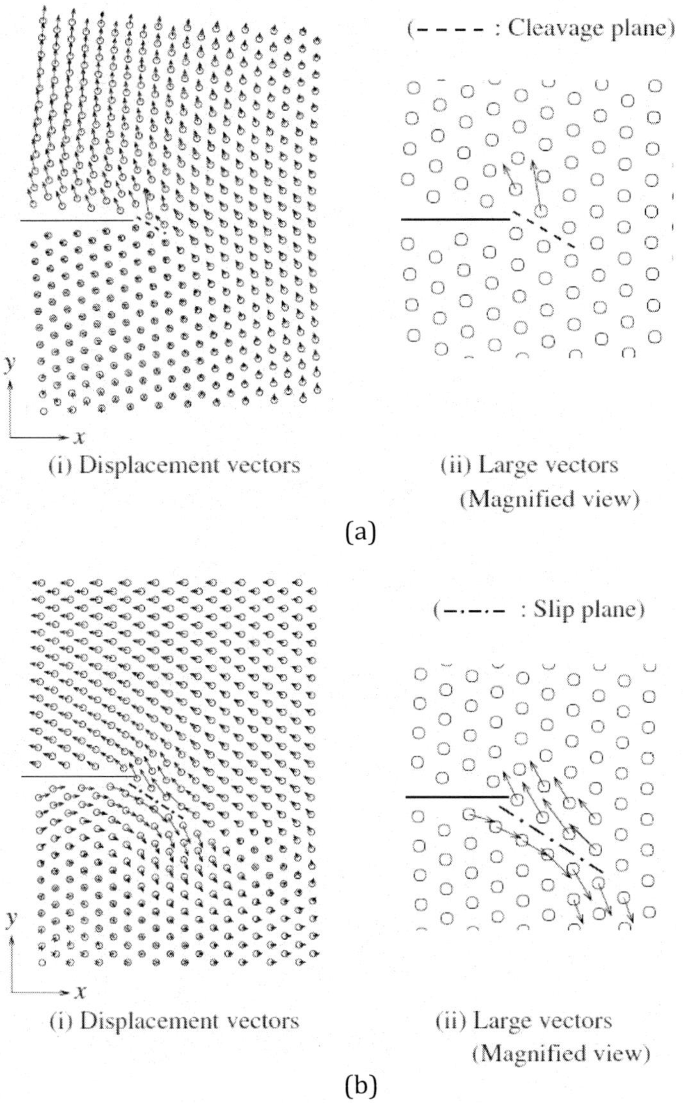

Figure 8.12 Unstable deformation modes (eigenvectors corresponding to minimum eigenvalues) for (a) Case A and (b) Case B. Reprinted from Kitamura, T., et al. (2004) "Analytical evaluation of unstable deformation criterion of atomic structure and its application to nanostructure," *Comp. Mater. Sci.*, 29, pp. 499-510, Copyright (2004), with permission from Elsevier.

8.2.2.2 Simplified evaluation of mechanical instability

In this section, we present a simplified method to evaluate mechanical instability in a large-scale atomic structure using a reasonable reduction of the degrees of freedom in the system.

In an attempt to reduce the degrees of freedom in the computation, we divide a part of the atomic system into elements and describe the displacement of atoms in the element using a function of the nodal coordinate. The displacement of atoms inside an element is calculated from the displacement of nodal atoms. For example, if we employ linear elements, the atomic displacement is linearly interpolated from that of nodal atoms. Figure 8.13 schematizes the concept applied to a two-dimensional structure of atoms with a crack. In the area far from the crack, the element includes many atoms because the strain gradient is moderate. On the other hand, the number of atoms in the element decreases in the region near the crack tip, and the freedom of each atom is taken into account in the area where the unstable deformation occurs preferentially.

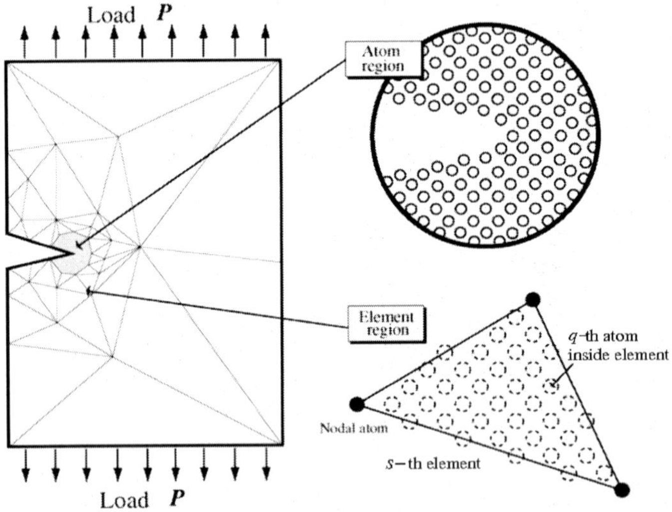

Figure 8.13 Schematic illustration explaining the concept of reducing degrees of freedom by using linear elements. Degrees of freedom of atoms inside an element are represented by its nodes.

Denoting the number of atoms and nodes in the system as N'_{atom} and N'_{node}, respectively, we can describe the deformation of the system by the vector $\mathbf{R}'$,

$$\mathbf{R}' \equiv^t (r_x^{\,1}, r_y^{\,1}, r_z^{\,1}, ..., r_i^{\,\alpha}, ..., r_x^{\,N'_{atom}}, r_y^{\,N'_{atom}}, r_z^{\,N'_{atom}},$$

$$S_x^{\,1}, S_y^{\,1}, S_z^{\,1}, ..., S_i^{\,\xi}, ..., S_x^{\,N'_{node}}, S_y^{\,N'_{node}}, S_z^{\,N'_{node}}) \tag{8.17}$$

where S_i^{ξ} denotes the i-th coordinate of a node, ξ. The number of degrees of freedom is $M' = 3(N'_{\text{atom}} + N'_{\text{node}}) - 6$, [$M' = 2(N'_{\text{atom}} + N'_{\text{node}}) - 3$ for a two-dimensional system]. The Hessian matrix, $\mathbf{A}'$, to be solved is then rewritten as

$$A'_{mn} = \frac{\partial^2 U}{\partial R'_m \partial R'_n} \tag{8.18}$$

where the size ($M' \times M'$) is significantly reduced from the original size ($M \times M$) because $M' << M$. Thus, by applying the concept, we can compute the eigenvalues (instability criterion) as well as the corresponding eigenvectors (unstable mode) even in a large-scale atomic structure.

Figure 8.14a shows the mesh division of a two-dimensional atomic component with a crack under the same displacement controlled condition as shown in Fig. 8.10b [Shimada et al., 2009a]. The atomic structure is divided into elements according to the strain gradient. Here, atomic displacement in an element is linearly related with the displacement at the nodes. The models of Cases A, B, and C have 250, 500, and 750 degrees of freedom, respectively. Note that the total number of degrees of freedom (atomic degrees of freedom) is 2524. The atomic region appears near the crack tip and elements become rough far from the crack. Figure 8.14b plots the minimum eigenvalues of the Hessian matrix for Cases A, B and C. The model with the lowest degrees of freedom (Case A) significantly differs from the exact numerical solution that considers all the degrees of freedom shown as "Precise." On the other hand, by increasing the degrees of freedom, Cases B and C give almost the same curves of the exact model, suggesting that this is a good approximation. Figure 8.15 shows the unstable eigenvector modes corresponding to the minimum eigenvalue for Cases A, B, and C. The unstable deformation is correctly reproduced by setting the

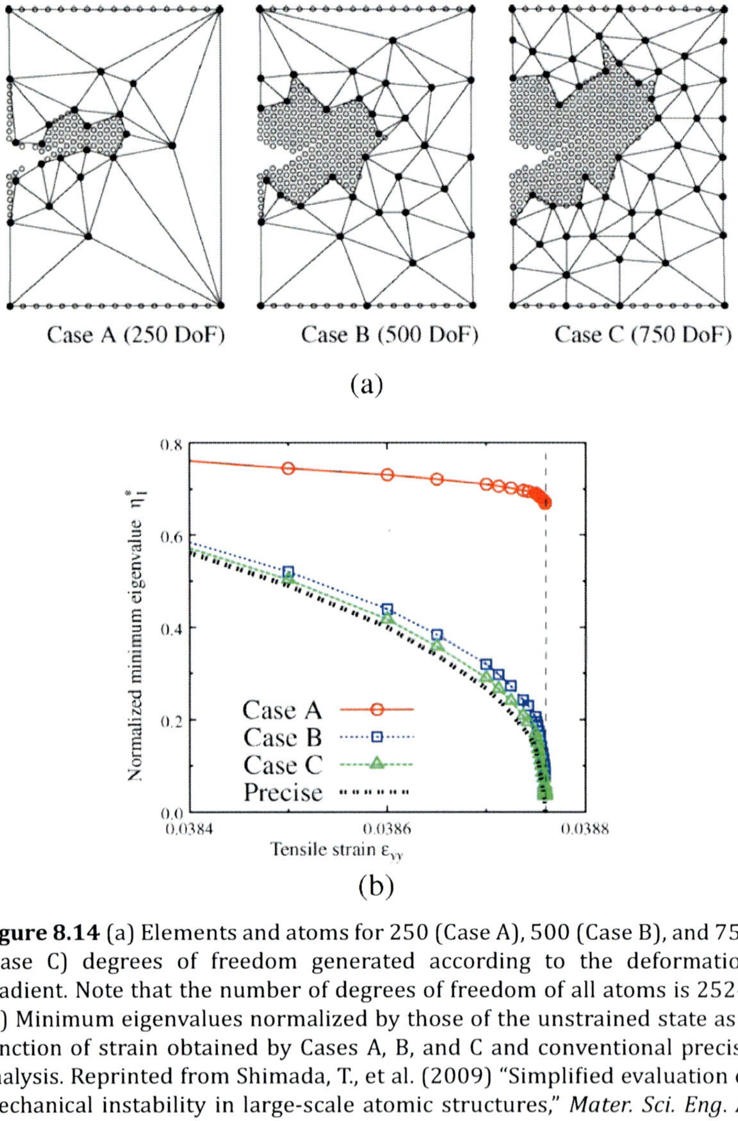

Figure 8.14 (a) Elements and atoms for 250 (Case A), 500 (Case B), and 750 (Case C) degrees of freedom generated according to the deformation gradient. Note that the number of degrees of freedom of all atoms is 2524. (b) Minimum eigenvalues normalized by those of the unstrained state as a function of strain obtained by Cases A, B, and C and conventional precise analysis. Reprinted from Shimada, T., et al. (2009) "Simplified evaluation of mechanical instability in large-scale atomic structures," *Mater. Sci. Eng. A*, 513–514, pp. 166–171, Copyright (2009), with permission from Elsevier.

degrees of freedom to more than 500 (Case B). Here, the size of matrix, **A**′, is reduced to about 1/25 compared with the original one, **A**. Although the model has a crystalline structure, it is also valid in highly disordered amorphous metals where the atomic displacement exhibits nonlinear behavior [Shimada et al., 2009b]. These examples are limited to two-dimensional degrees of freedom. However, in recent years, we have applied the method to a three-dimensional atomic component with a nano-scale crack, as shown in Fig. 8.16. Thus, there is no limitation to the simplified analysis method [Shimada et al., 2010].

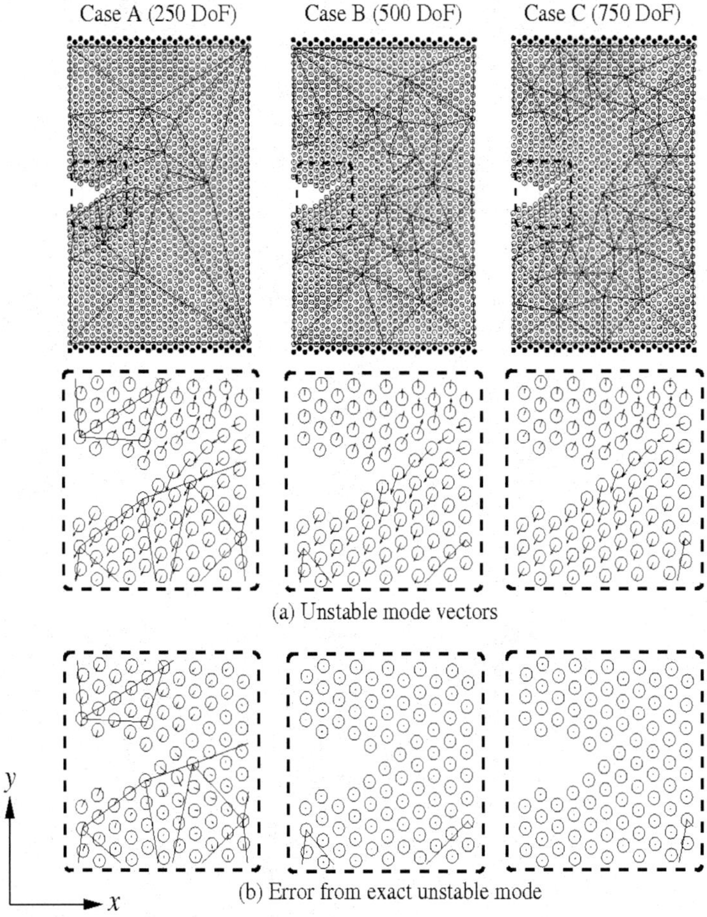

Figure 8.15 Simplified instability analysis results for Cases A, B, and C. (a) Unstable mode vectors. (b) Error vectors representing the difference from the exact unstable mode. Only the region near the crack tip is shown. Reprinted from Shimada, T., et al. (2009) "Simplified evaluation of mechanical instability in large-scale atomic structures," *Mater. Sci. Eng. A*, 513 -514, pp. 166 -171, Copyright (2009), with permission from Elsevier.

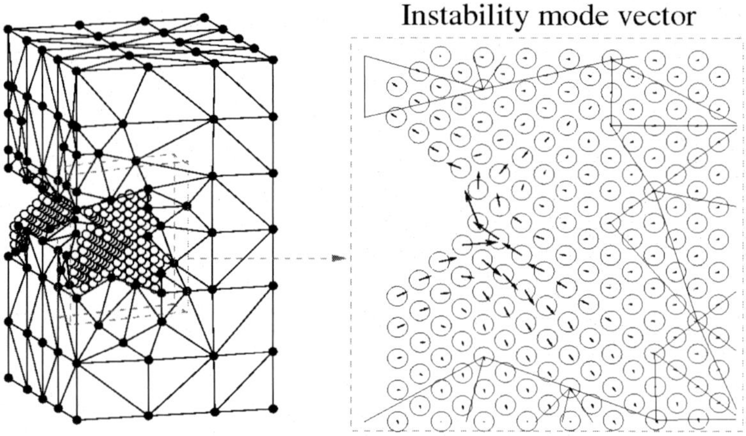

Figure 8.16 An example for three-dimensional result of simplified instability analysis: Application to a nano-scale crack in a three-dimensional atomic component. Reprinted with permission from Shimada, T., et al. (2010) "Simplified analysis of mechanical instability in three-dimensional atomic components and its application to nanoscale crack," *Int. Solid Mech. Mater. Eng.*, 4, pp.1071–1082.

References

Baskes, M. I., Angelo, J. E., and Bisson, C. L. (1994) Atomic calculations of composite interfaces, *Mod. Simul. Mater. Sci. Eng.*, 2, pp. 505–518.

Becker, T. L., McNaney, J. M., Cannon, R. M., and Ritchie, R. O. (1997) Limitation on the use of mixed-mode delaminating beam test specimen: effects of the size of the region of K-dominance, *Mech. Mater.*, 25, pp. 291–308.

Born, M., and Huang, K. (1954) *Dynamical Theory of Crystal Lattices*, Oxford, Oxford University Press.

Celis, B. D., Argon, A. S., and Yip, S. (1983) Molecular dynamics simulation of crack tip process in alpha-iron and copper, *J. Appl. Phys.*, 54, pp. 4864–4878.

Cheung, K. S., and Yip, S. (1990) Brittle-ductile transition in intrinsic fracture behavior of crystals, *Phys. Rev. Lett.*, 65, pp. 2804–2807.

Clatterbuck, D. M., Krenn, C. R., Cohen, M. L., and Morris, Jr. (2003) Phonon instabilities and the ideal strength of aluminum, *Phys. Rev Lett.*, 91, article number 135501.

Dmitriev, S. V., Kitamura, T., Li, J., Umeno, Y., Yashiro, K., and Yoshikawa, N. (2005) Near-surface lattice instability in 2D fiber and half-space, *Acta Mater.*, 53, pp. 1215–1224.

Gall, K., Horstemeyer, M. F., Schilfgaarde, M. V., and Baskes, M. I. (2000) Atomistic simulations on the tensile debonding of an aluminum-silicon interface, *J. Mech. Phys. Solids*, 48, pp. 2183–2212.

Girifalco, L. A., and Weizer, G. V. (1959) Application of the Morse potential function to cubic metals, *Phys. Rev.*, 114, pp. 687–690.

Hunter, L. (1942) The variation with temperature of the principal elastic moduli of NaCl near the melting point, *Phys. Rev.*, 61, pp. 84–90.

Kitamura, T., Hirakata, H., and Satake, Y. (2004a) Applicability of fracture mechanics on brittle delamination of nano-scale film edge, *JSME Int. J. Ser. A*, 47, pp. 106–112.

Kitamura, T., Umeno, Y., and Tsuji, N. (2004b) Analytical evaluation of unstable deformation criterion of atomic structure and its application to nanostructure, *Comp. Mater. Sci.*, 29, pp. 499–510.

Kitamura, T., Umeno, Y., and Fushino, R. (2004c) Instability criterion of inhomogeneous atomic system, *Mater. Sci. Eng. A*, 379, pp. 229–233.

Kitamura, T., Hirakata, H., and Itsuji, T. (2003) Effect of residual stress on delamination from interface edge between nano-films, *Eng. Frac. Mech.*, 70, pp. 2089–2101.

Kubo, S., Nakata, M., Ohji, K., and Ishii, A. (1999) Analyses of deformation and fracture near interface crack tip using molecular dynamics method, *Trans. JSME*, 65, pp. 560–566, (in Japanese).

Milstein, F., and Farber, B. (1980) Theoretical fcc→bcc transition under [100] tensile loading, *Phys. Rev. Lett.*, 44, pp. 277–280.

Mullins, M., and Dokainish, M. A. (1982) Simulation of the (001) plane crack in α-iron employing a new boundary scheme, *Philos. Mag. A*, 46, pp. 771–787.

Paskin, A., Gohar, A., and Dienes, G. J. (1980) Computer simulation of crack propagation, *Phys. Rev. Lett.*, 44, pp. 940–943.

Shimada, T., Okawa, S., Minami, S., and Kitamura, T. (2009a) Simplified evaluation of mechanical instability in large-scale atomic structures, *Mater. Sci. Eng. A*, 513–514, pp. 166–171.

Shimada, T., Okawa, S., Minami, S., and Kitamura, T. (2009b) Development of efficient instability analysis method for atomic structures using linear elements and its application to amorphous metal, *J. Solid Mech. Mater. Eng.*, 3, pp. 807–818.

Shimada, T., Okawa, S., and Kitamura T. (2010) Simplified analysis of mechanical instability in three-dimensional atomic components and its application to nanoscale crack, *Int. Solid Mech. Mater. Eng.*, 4, pp.1071–1082.

Suga, T., Funahashi, A., and Takahashi, Y. (1991) A molecular dynamics approach to interface crack problem, *Trans. JSME*, 57, pp. 678–685, (in Japanese).

Tse, J. S., and Klug, D. D. (1991) Mechanical instability of α-quartz: a molecular dynamics study, *Phys. Rev. Lett.*, 67, pp. 3559–3562.

Umeno, Y., Kitamura, T., and Tagawa, M. (2007) Mechanical instability in non-uniform atomic structure: application to amorphous metal, *Mater. Sci. Eng. A*, 462, pp. 450–455.

Van Vliet, K. J., Li, J., Zhu, T., Yip, S., and Suresh S. (2003) Quantifying the early stages of plasticity through nano-scale experiments and simulations, *Phys. Rev. B*, 67, article number 104105.

Wang, J., Yip, S., Phillpot, S. R., and Wolf, D. (1993) Crystal instabilities at finite strain, *Phys. Rev. Lett.*, 71, pp. 4182–4185.

Yashiro, K., and Tomita, Y. (2001) Local lattice instability at a dislocation nucleation and motion, *J. Phys. IV*, 11, pp. Pr5-3–Pr5-10.

Index